Advanced Tactical Fighter to F-22 Raptor: Origins of the 21st Century Air Dominance Fighter

Advanced Tactical Fighter to F-22 Raptor: Origins of the 21st Century Air Dominance Fighter

David C. Aronstein
Michael J. Hirschberg
Albert C. Piccirillo
ANSER
Arlington, Virginia

American Institute of Aeronautics and Astronautics, Inc.
1801 Alexander Bell Drive
Reston, VA 20191

Publishers since 1930

American Institute of Aeronautics and Astronautics, Inc., Reston, Virginia

1 2 3 4 5

Library of Congress Cataloging-in-Publication Data

Aronstein, David C.
    Advanced tactical fighter to F-22 raptor : origins of the 21st
century air dominance fighter / David C. Aronstein, Michael J.
Hirschberg, Albert C. Piccirillo.
        p.   cm.
    Includes bibliographical references and index.
    ISBN 1-56347-282-1 (softcover : alk. paper)
    1. F-22 (Jet fighter plane)—Design and construction.
I. Hirschberg, Michael J.   II. Piccirillo, Albert C.   III. Title.
TL685.3.A76   1998   623.7′464—dc21   98-27265

Cover design by Sara Bluestone

ACKNOWLEDGMENTS

The authors would like to express their sincere appreciation to the following for their strong interest and support in this endeavor: Lt. Gen. George K. Muellner, U.S. Air Force, deputy assistant secretary of the Air Force for Acquisition; John J. Welch, assistant secretary of the Air Force for Acquisition from 1987 to 1991; and Sherman N. Mullin, Lockheed Advanced Tactical Fighter (ATF) and F-22 program manager from 1986 to 1991 and then president of the Lockheed Advanced Development Company (the Skunk Works) from 1991 to 1994.

Delbert Jacobs (Brigadier General, U.S. Air Force, retired), former Northrop ATF Program Manager and a key player in the development of the YF-23, provided many useful insights. John Cashen, former head of stealth programs at Northrop, provided information on the evolution of stealth technologies at Northrop; Alan Brown, formerly head of stealth technology efforts at Lockheed's Skunk Works, gave insights into Lockheed's approaches. Col. Robert F. Lyons, U.S. Air Force, and Lt. Col. Charles W. Pinney, U.S. Air Force, both former key players in F-22 avionics development, added valuable inputs to this aspect of the ATF/F-22 story.

Mr. Wayne O'Connor of the U.S. Air Force, Aeronautical Systems Center (ASC) contributed insights on in-house U.S. Air Force design analysis efforts. Extensive material on the ATF engine development effort was provided by David A. Morris and James D. Smith, both of General Electric, and by Gary A. Plourde, the Pratt and Whitney ATFE program manager (from 1981 to 1989), and Paul V. Davisson, also of Pratt and Whitney.

An especially noteworthy contribution was provided by Paul C. Ferguson, staff historian at the ASC History Office at Wright–Patterson Air Force Base, Ohio. Dr. Ferguson archived valuable documents related to the ATF/F-22 Program; he is developing an in-depth phase-by-phase history that focuses on the ASC's role in the program. Diana G. Cornelisse, chief of the ASC History Office, cheerfully went out of her way to assist in our research.

Eric Hehs, managing editor of Lockheed Martin's *Code One Magazine*, provided a draft of his forthcoming article on the early days of the ATF program as well as valuable material on the Dem/Val designs proposed to the U.S. Air Force by the members of the current Lockheed Martin team. Both Lockheed Martin and Northrop Grumman provided the majority of the photos of the YF-22 and YF-23 aircraft. Jack Butler and Dimitri Savvas of ANSER assisted with the graphics. Finally, we would like to thank Heather Brennan and Megan Scheidt of the AIAA Books Program, who served as editors for this project.

David C. Aronstein is a principal aerospace engineer at ANSER, where he provides senior technical support to several aerospace programs including the Joint Strike Fighter and the F-22 air superiority fighter. Previously he was employed at Boeing Commercial Airplane Company on the 777 and High Speed Commercial Transport programs in subsonic and supersonic aerodynamic design and analysis. Dr. Aronstein also has performed aircraft structural analysis and experimental fluid dynamics research related to the internal cooling flows in rotating turbine engine components. He holds a B.S. in mechanical and aerospace engineering with highest honors from Princeton University, an M.S. in aeronautics and astronautics from Stanford University, and a Ph.D. in aeronautics and astronautics from the University of Washington. Dr. Aronstein is a Senior Member of the American Institute of Aeronautics and Astronautics.

Michael J. Hirschberg is an aerospace engineer at ANSER. He currently supports the Propulsion Management Team in the Joint Strike Fighter Program Office. Previous positions included supporting the Fighter Division within the Office of the Assistant Secretary of the Air Force for Acquisition on the F-22 Advanced Tactical Fighter and F119 engine programs and working as a propulsion engineer on various solid rocket motor development programs with Aerojet General and the Naval Air Systems Command. Mr. Hirschberg received his B.S. in aerospace engineering from the University of Virginia and his M.S. in mechanical engineering from the Catholic University of America. He is a Senior Member of the American Institute of Aeronautics and Astronautics.

Albert C. Piccirillo (Colonel, U.S. Air Force, retired) is manager of the Joint Technology Division at ANSER. While assigned to the U.S. Air Force Aeronautical Systems Division from 1983 to 1987, he served as Advanced Tactical Fighter (ATF) system program director. Col. Piccirillo headed the Source Selection Board that resulted in the award of contracts to Lockheed and Northrop for development and flight test of the YF-22 and YF-23 air superiority fighter prototypes. He received an *Aviation Week and Space Technology* Laurel Award in 1986 for "setting, explaining and meeting the goals of the new fighter development program." Previously, as chief of the Tactical Systems Division in the Air Force Center for Studies and Analyses, he was responsible for analyzing advanced aircraft, weapons, and employment concepts. At the Air

Force Flight Test Center, he led an engineering team responsible for flight test of unmanned aerial vehicles. He is an Air Force command pilot with 3000 hours of flying time, including combat assignments flying F-4 fighters in Southeast Asia where he flew 200 combat missions. Mr. Piccirillo has a B.S. in aeronautical engineering from Pennsylvania State University and an M.S. in aerospace engineering from the Air Force Institute of Technology. He completed the program manager course at the Defense Systems Management College and is an Associate Fellow of the American Institute of Aeronautics and Astronautics.

TABLE OF CONTENTS

LIST OF ACRONYMS AND ABBREVIATIONS

A/F-X	Attack/Fighter Experimental
A-A	air-to-air
AAAM	advanced air-to-air missile
AASF	Advanced Air Superiority Fighter
ACA	associate contractor agreement
ACCD	advanced medium range air-to-air missile compressed carriage demonstration
ACEMA	advanced counterair engagement mission analysis
ACF	Air Combat Fighter
ADM	advanced development model
ADTC	Armament Development Test Center
AEDC	Arnold Engineering Development Center
AF CAIG	Air Force Cost Analysis Improvement Group
AF/SA	Air Force Center for Studies and Analyses
AFAL	Air Force Avionics Laboratory
AFATL	Air Force Armament Laboratory
AFFDL	Air Force Flight Dynamics Laboratory
AFFTC	Air Force Flight Test Center
AFL	avionics flying laboratory
AFOTEC	Air Force Operational Test and Evaluation Center
AFSARC	Air Force Systems Acquisition Review Council
AFSC	Air Force Systems Command
AFTI	Advanced Fighter Technology Integration
AFWAL	Air Force Wright Aeronautical Laboratories
A-G	air-to-ground
AGP	avionics ground prototype
AGT	advanced gun technology
AIM	air intercept missile
AIWS	advanced interdiction weapon system
AMRAAM	advanced medium range air-to-air missile
AMT	accelerated mission testing
AMTA	Advanced Manned Tactical Aircraft
ANSER	Analytic Services, Inc.
AOA	angle of attack
APB	acquisition program baseline
APL	Aero Propulsion Laboratory
APSI	Aircraft Propulsion Subsystem Integration
ARPA	Advanced Research Projects Agency
ASD	Aeronautical Systems Division (of AFSC)
ASD/RW	Aeronautical Systems Division Deputy for Electronic Combat and Reconnaissance
ASD/RWW	Aeronautical Systems Division Electronic Warfare System Program Office

ASD/TA	Aeronautical Systems Division Deputy for Tactical Systems
ASD/XR	Aeronautical Systems Division Deputy for Development Planning
ASD/YF	Advanced Tactical Fighter System Program Office (after it left the Deputy for Tactical Systems)
ASD/YZ	Aeronautical Systems Division New Engine System Program Office
ASRAAM	advanced short range air-to-air missile
ASTOVL	advanced short takeoff and vertical landing
ASW	anti-submarine warfare
ATA	Advanced Tactical Aircraft
ATAS	advanced tactical attack system
ATASMA	advanced tactical attack system mission analysis
ATEGG	Advanced Turbine Engine Gas Generator
ATES	advanced technology engine studies
ATF	Advanced Tactical Fighter
ATFE	Advanced Tactical Fighter Engine
ATFMA	Advanced Tactical Fighter mission analysis
ATGAF	Air-to-Ground Advanced Fighter
ATS	air-to-surface
ATSF	Advanced Tactical Strike Fighter
AWACS	airborne warning and control system
A-X	Attack Experimental
BAFO	best and final offer
BES	budget estimate submission
BIT	built-in-test
BLATS	built-up low-cost advanced titanium structure
BMI	bismaleimide
BPR	bypass ratio
C&A	controls and accessories
CAB	common avionics baseline
CAP	Combat Aircraft Prototype
CAP	common avionics processor
CAS	close air support
CAS/BI	close air support/battlefield interdiction
CAT	combat aircraft technology
CBO	Congressional Budget Office
CCV	control configured vehicle
CD	concept definition
CDI	concept development investigation
CDR	critical design review
CDT	concept development team
CET	combustor exit temperature
CFE	contractor furnished equipment
CISPO	Combat Identification Systems Program Office
CMP	common module program

CNI	communication, navigation, and identification
COEA	cost and operational effectiveness analysis
COMSEC	communications security
CONOPS	concept of operations
CONUS	Continental United States
CPAF	cost plus award fee
CRF	Compressor Research Facility (Wright Laboratories at Wright–Patterson Air Force Base)
CTD	critical technology demonstration
CTF	combined test force
DAB	Defense Acquisition Board
DARPA	Defense Advanced Research Projects Agency
DDR	detailed design review
Dem/Val	demonstration and validation
DIA	Defense Intelligence Agency
DOD	Department of Defense
DRB	Defense Resources Board
DSARC	Defense Systems Acquisition Review Council
EC	electronic combat
ECCM	electronic counter-countermeasures
ECM	electronic countermeasures
EMD	engineering and manufacturing development
EO	electro-optical
EOA	early operational assessment
ETF	Enhanced Tactical Fighter
EW	electronic warfare
FADEC	full authority digital electronic control
FETT	first engine to test
FFAS	Future Fighter Alternatives Study
FFP	firm fixed price
FLAMR	forward looking advanced multimode radar
FOD	foreign object damage
FOG	finger-on-glass
FPIF	fixed price plus incentive fee
FSD	full-scale development
FYDP	five-year defense plan
GAO	General Accounting Office
GFE	government furnished equipment
GPS	Global Positioning System
GRC	General Research Corporation

HARM	high-speed anti-radiation missile
HASC	House Armed Services Committee
HI	"Hi-Mach/Hi-Altitude"
HiMAT	Highly Maneuverable Advanced Technology
HQ USAF	Air Force Headquarters
HUD	headup display
IADS	integrated air defense system
ICG	Interface Control Group
ICNIA	Integrated Communication, Navigation, and Identification Avionics
IFF	identification friend or foe
IHPTET	Integrated High Performance Turbine Engine Technology
INEWS	Integrated Electronic Warfare System
IOC	initial operational capability
IOT&E	initial operational test and evaluation
IR	infrared
IR&D	internal (i.e., company-funded as opposed to contracted) research and development
IRCCM	infrared counter-countermeasures
IRST	infrared search and track
ITAS	improved tactical attack system
JAFE	Joint Advanced Fighter Engine
JFE	Joint Fighter Engine
JHMCS	Joint Helmet Mounted Cueing System
JIAWG	Joint Integrated Avionics Working Group
JRMB	Joint Requirements Management Board
JSF	Joint Strike Fighter
JTDE	Joint Technology Demonstrator Engine
JTIDS	Joint Tactical Information Distribution System
KEAS	knots equivalent airspeed
KIAS	knots indicated airspeed
LANTIRN	low-altitude navigation and targeting infrared night
LCC	life cycle cost
LHX	Light Helicopter Experimental
LO	low observable
LPI	low probability of intercept
LRIP	low rate initial production
M/S	milestone
MAR	major aircraft review
MENS	mission element need statement
MFD	multifunction display
MLS/ILS	microwave landing system/instrument landing system

MOA	memorandum of agreement
MOU	memorandum of understanding
MR	modification request
MRC	major regional conflict
MTBF	mean time between failures
MTI	moving target indication
MTOW	maximum takeoff weight
NATF	Navy Advanced Tactical Fighter
NATO	North Atlantic Treaty Organization
NAVAIR	Naval Air Systems Command
NBC	nuclear/biological/chemical
NFA	new fighter aircraft
O&S	operations and support
OASMA	Offensive Air Support Mission Analysis
OFP	operational flight program
OMB	Office of Management and Budget
OPR	office of primary responsibility
OPR	overall (engine) pressure ratio
OSD	Office of the Secretary of Defense
OT&E	operational test and evaluation
OUSDR&E	Office of the Undersecretary of Defense (Research and Engineering)
P3I	preplanned product improvement
PATS	propulsion assessment for tactical systems
PAV	prototype air vehicle
PMD	program management directive
POM	program objective memorandum
PSC	preferred system concept
PSOC	preliminary system operational concept
PSS	preliminary system specification
R&M	reliability and maintainability
RAM	radar absorbent material
RCS	radar cross section
RD	office symbol of the Deputy Chief of Staff for Research and Development, Headquarters U.S. Air Force
RDQT	Air Force Headquarters, Deputy for Research and Development, Tactical Division (this was the office of primary responsibility within Headquarters U.S. Air Force for the Advanced Tactical Fighter during the early stages of the program)
RDT&E	research, development, test, and evaluation
RF	radio frequency
RFI	request for information

RFP	request for proposal
RM&S	reliability, maintainability, and supportability
ROC	required operational capability
RSR	rapid solidification rate
S/MTD	short takeoff and landing/Maneuver Technology Demonstrator
S^3	Strike System Study (later changed to Synergistic Strike System)
SA	Air Force studies and analyses
SAB	Scientific Advisory Board
SAE	Society of Automotive Engineers
SAF	Secretary of the Air Force
SAF/AL	Assistant Secretary of the Air Force (Acquisition and Logistics) (old)
SAF/AQ	Assistant Secretary of the Air Force (Acquisition) (current)
SAF/AQPF	Fighter Division within the Office of Assistant Secretary of the Air Force (Acquisition)
SAM	surface-to-air missile
SAR	special access required
SAR	synthetic aperture radar
SASC	Senate Armed Services Committee
SCAMP	Supersonic Cruise and Maneuver Prototype
SCM	"Supersonic Cruise and Maneuver"
SCP	system concept paper
SDR	system design review
SEAD	suppression of enemy air defenses
SECDEF	Secretary of Defense
SEM-E	standard electronics model E
SFC	specific fuel consumption
SGR	sortie generation rate
SICBM	small intercontinental ballistic missile
SLO	"subsonic low observable"
SMS	stores management system
SOC	system operational concept
SON	statement of need
SORD	system operational requirements document
SPO	System Program Office
SRAM II	Short Range Attack Missile II
SRR	system requirements review
SSD	system specification development
SSP	source selection plan
STAR	System Threat Analysis Report
STOL	short takeoff and landing
SUWACS	Soviet Union Airborne Warning and Control System
T/R	transmit-receive
T/W	thrust-to-weight

TAC	Tactical Air Command
TAC	total accumulated cycle
TAC/CC	Commander, Tactical Air Command
TAC/DR	Director of Requirements, Tactical Air Command
TACAIR	tactical air power
TACAN	tactical air navigation
TAD	technology availability date
TAF SON	Tactical Air Forces Statement of Operational Need
TAFTA	tactical fighter technology alternatives
TASS	technical analytical study support
TAWD	target acquisition and weapon delivery
TBC	thermal barrier coating
TOGW	takeoff gross weight
TY	then-year
UAV	unmanned aerial vehicle
UFC	unit flyaway cost
UHF	ultra high frequency (radio)
UPC	unit procurement cost
URR	Ultra Reliable Radar
USAF/CV	vice chief of staff of the U.S. Air Force
USAF/RDQ	Directorate of Operational Requirements, under the deputy chief of staff for Research and Development, U.S. Air Force Headquarters
USDR&E	Undersecretary of Defense (Research and Engineering)
V/STOL	vertical/short takeoff and landing
VABI	variable area bypass injector
VCE	variable cycle engine
VHF	very high frequency (radio)
VHSIC	very high-speed integrated circuit
VLO	very low observable
VMS	vehicle management system
WL	Wright Laboratories
WSC	weapon systems contractor
XF	in the context of the Advanced Tactical Fighter Engine program, the ground demonstrator engines
XRH	Directorate for Design Analysis
XRJ	Directorate of Low Observables
XRM	Directorate of Mission Analysis
XST	experimental survivable testbed
YF	in the context of the Advanced Tactical Fighter Engine program, the demonstration and validation flight test engines
YF-	in the context of the Advanced Tactical Fighter program, the prefix for the demonstration and validation prototype aircraft

Chapter 1

INTRODUCTION

The F-22 is intended to be the United States's front-line air superiority fighter from its planned initial operational capability (IOC) in 2005 through the first quarter of the 21st century. Its overall objective can be described as providing "air dominance," i.e., the ability to not only control all friendly airspace but to dominate hostile airspace at any time and place of the U.S./Allied theater commander's choosing, in any type of conflict, against any adversary.[1]

The F-22 will achieve this goal through a synergistic combination of characteristics, at the center of which are 1) stealth; 2) supersonic cruise speeds, sustained without the use of afterburners (supercruise); and 3) integrated avionics. Additional features of the F-22 are superior maneuverability; increased range relative to prior-generation fighter aircraft; high lethality; improved reliability, maintainability, and supportability (RM&S); and a secondary precision air-to-ground capability.

The Lockheed Martin/Boeing F-22, powered by Pratt and Whitney F119 engines, is the result of the Advanced Tactical Fighter (ATF) program. The ATF program began in the early 1970s, originally with air-to-ground as the primary role. Air-to-air missions began to be considered during the late 1970s, and since 1982 the consistent aim of the ATF program has been to provide what is now referred to as "air dominance." During this period technologies have been identified, matured, demonstrated, and transitioned; airframe and engine designs have been developed and demonstrated; and requirements have been carefully refined to provide the necessary mission capabilities as cost effectively as possible.

The concept definition (CD) phase of the ATF program formally began in November 1981. Demonstration and validation (Dem/Val) was initiated in October 1986, and engineering and manufacturing development (EMD) began in August 1991. This document traces the history of the ATF program and the evolution of ATF requirements from the beginning of the program approximately through the start of EMD. The competing ATF design, the Northrop/McDonnell Douglas YF-23, and the competing General Electric YF120 engine are also discussed in the context of the ATF program. Because the YF120 is now being developed into an alternate engine for the Joint Strike Fighter (JSF), additional detail is provided on it.

During EMD, development of the F-22 continues; several major rephasings (due to funding cuts), as well as perceptions that the threat has lessened due to the dissolution of the Soviet Union, have slowed the program considerably. The original projected total production quantity has also been reduced from 750 aircraft, which was the basis of EMD planning, to the current 339. The preliminary design review for the EMD F-22A was completed on April 30, 1993, with the critical design review being completed on February 24, 1995. Fabrication of the first part of the EMD F-22, a titanium section of the aft fuselage, began on December 8, 1993; assembly began on June 27, 1995, when the first mid-fuselage bulkhead was loaded into an assembly fixture. On April 9, 1997, the EMD F-22 was publicly unveiled for the first time and dubbed the Raptor. First flight of the EMD F-22 was conducted by Chief Test Pilot Paul Metz on September 7, 1997 (Fig. 1). Flight testing (involving both developmental and initial operational test and evaluation) using nine EMD aircraft will continue for a number of years, with the F-22A Raptor currently scheduled to enter operational service with the U.S. Air Force Air Combat Command at the end of the year 2004.

Fig. 1   First flight of the Lockheed Martin F-22 Raptor, the U.S. Air Force's Air Dominance Fighter for the 21st century, occurred on September 7, 1997, at Marietta, Georgia.

REFERENCE

[1]Operational Requirements Document (U) CAF 304-83-I/II-A (Rev. 2), *F-22 Advanced Tactical Fighter,* HQ ACC/DR SMO-22, Aug. 26, 1996, with Attachment: *Requirements Correlation Matrix.* SECRET. (Unclassified information only used from this source.)

Chapter 2

BACKGROUND: THE ADVANCED TACTICAL FIGHTER PRIOR TO MILESTONE ZERO

OVERVIEW

The concept of an Advanced Tactical Fighter (ATF) was first proposed during the early 1970s as an advanced *air-to-surface* strike aircraft to replace the McDonnell Douglas F-4, the Fairchild Republic F-105, and/or the General Dynamics F-111 in front-line U.S. Air Force service. (These aircraft, developed in the 1950s and early 1960s, are illustrated in Fig. 2.) Throughout the 1970s, various requirements documents, design studies, technology studies, mission area analyses, and related efforts were directed at defining the ATF and preparing for its development. The first formal direction to address air-to-air mission capabilities for the ATF came in 1980. This led to the consideration of a set of options that included a pure air-to-surface strike fighter, a pure air superiority fighter, two separate aircraft (one for each role), or a single multimission aircraft.

From 1980 through 1982, the primary emphasis shifted toward the air superiority role. There were several reasons for this shift. When ATF studies first began to address both mission areas, it became apparent that air-to-air capability was more of a design driver than air-to-ground. This discovery simply confirmed the historical pattern that air superiority fighters can often be adapted to perform air-to-surface missions, but not the reverse. Furthermore, new developments in the Soviet air-to-air threat indicated an impending shortfall in this area, while the introduction of various improved U.S. air-to-surface capabilities (including the F-117A) made the need for another new U.S. strike aircraft less urgent. During the same period (in November 1981), the ATF program received Milestone Zero approval, marking its initiation as a formal weapon system acquisition program. However, fiscal year 1982 funding for the ATF program was denied, resulting in a lapse of approximately one year between Milestone Zero approval and the initiation of funded concept definition activities.

Fig. 2   Until the early 1980s, the ATF was conceived as an air-to-surface strike aircraft. It was originally intended as a replacement for the F-4 (top), the F-105 (middle), and possibly the F-111 (bottom).

AIR-TO-SURFACE ADVANCED TACTICAL FIGHTER

Although the ATF concept centered on the air-to-surface role for the entire decade of the 1970s, many important weapon system attributes, which characterize the ATF as it finally took shape, were identified during this period. These include 1) increased supersonic persistence relative to prior-generation aircraft, 2) integrated avionics, 3) reduced observables, and 4) improved supportability. This section therefore describes the early evolution of the ATF concept.

U.S. AIR FORCE TACTICAL FORCES 1985 STUDY

This study, conducted by the Tactical Air Command (TAC) during 1969 and 1970, is commonly identified as the origin of the Advanced Tactical Fighter concept. The objectives of the Tactical Forces 1985 Study (TAC-85) were to

1) forecast the environment that tactical Air Forces would have to operate in during the period from 1970 to 1985 and
2) define capabilities needed by the tactical Air Forces during that period, in the areas of command and control, reconnaissance, special air forces, airlift, and fighters.

For perspective, it should be noted that the F-15 was still in development, and the Lightweight Fighter experimental prototyping program (which eventually led to the F-16) had not yet begun. The Vietnam War was ongoing; lessons from that war were in the process of being learned. One of these lessons was that ground-based air defense systems, comprising networked radars, surface-to-air missiles (SAMs), and radar-guided anti-aircraft guns, had become a serious threat to all tactical aircraft and would have to be considered in any future planning. The TAC-85 study underscored the need for a highly survivable tactical strike aircraft for mid-to-high intensity conflicts during the time period addressed by the study.[1,2]

TACTICAL AIR COMMAND CONCEPT OF OPERATIONS

TAC issued a concept of operations (CONOPS) for such an aircraft, identified as the Advanced Tactical Fighter (also referred to as the Advanced Offensive Strike Fighter), in January 1971. The CONOPS identified three primary mission areas associated with an offensive air campaign: 1) air interdiction, 2) close air support (CAS), and 3) offensive and defensive air-to-air.

The McDonnell Douglas F-15 and the Attack-Experimental (A-X) aircraft, seen in Fig. 3 (the latter to be introduced as the Fairchild A-10), both still under development at the time of this first ATF CONOPS, were considered capable of meeting the expected needs in the air-to-air and close air support mission areas, respectively. This left the most pressing need in the air interdiction mission area:

Fig. 3 The F-15 was developed as a dedicated air superiority fighter. The A-10 emerged from the A-X competition as the dedicated U.S. Air Force close air support aircraft.

> Currently, this mission is assigned to the F-4 and F-111 fleet which is rapidly becoming obsolete in face of the enemy's advanced air defense aircraft and ground defense systems. In the post 1980 time period, this mission will be allocated to the Advanced Offensive Strike Fighter [i.e., the Advanced Tactical Fighter].[3]

In addition, the ATF was envisioned to have a secondary capability to perform close air support. However, the ATF's only air-to-air capability would be for self-defense.

The 1971 ATF CONOPS placed strong emphasis on the need for survivable penetration, expected to be achieved through high speed, maneuverability, electronic countermeasures (ECM), and situational awareness [airborne warning and control system (AWACS) as well as onboard threat warning systems]. Another area of emphasis was on targeting and weapons delivery, in particular at night and/or in adverse weather.

FISCAL YEAR 1972 PARAMETRIC AND POINT DESIGN STUDIES

In April and June of 1971, representatives from TAC, Air Force Headquarters (HQ USAF), Air Force Systems Command (AFSC) Headquarters, the Aeronautical Systems Division (ASD) of AFSC, and Analytic Services, Inc. (ANSER) met to "identify near term actions which have to be taken to further the concept of an ATF for TAC."[4] Following these meetings, AFSC tasked ASD to obtain preliminary design tradeoff analyses for an ATF. Eight aircraft companies submitted proposals in November 1971 to perform the study. Contracts of just under $200,000 each were awarded to General Dynamics and McDonnell Douglas to perform two tasks:

1) Technology assessment to identify opportunities for improved performance, improved mission effectiveness, and/or reduced cost. Technologies to be available for engineering development by 1976.
2) Parametric and point design tradeoff analyses "needed to make an initial selection of the aircraft and the mission subsystems characteristics necessary for an improved strike capability in the future threat environments."[1]

The analysis looked at aircraft designs over a broad range of speed/altitude combinations, including attack missions flown at supersonic speeds. The studies were conducted from February through July 1972. Results of the studies, together with an internal ASD analysis, were then made available to TAC for use in drafting a required operational capability (ROC) document.[1]

The first formal ATF requirements document, TAC ROC 301-73, was issued in draft form on January 26, 1973. At that time, a ROC was the primary document used to identify an operational need and request the development of a new capability to meet such a need. The initial version of the ROC reportedly was "written around a high-subsonic aircraft operating at medium altitudes."[5] The draft ROC was circulated to ASD, Air Staff, and other Air Force agencies for comment during 1973. Comments indicated a need for better definition of navigation, fire control, and weapon delivery accuracy requirements, and for efforts to improve survivability through ECM and aircraft radar cross section reduction.[1]

Concurrently, internal Air Force analyses of ATF designs, technologies, and missions were conducted. However, attempts to step up the level of activity (including an ATF prototype effort proposed to begin in FY75) were unsuccessful. There appeared to be several outstanding issues concerning an ATF program:

1) There was not a consensus that the ROC, as written, justified the development of a new aircraft. There was some feeling that the mission could be accomplished by the F-4, the F-15, or the F-111.
2) Alternatives, including not only existing and programmed aircraft but also remotely piloted vehicles or standoff missiles, needed to be investigated.
3) There appeared to be an inconsistency in the ROC between the need for a low-cost ATF and the sophisticated avionics and weapon delivery capabilities that were required.

As a result, the ROC as first released was not validated during 1973.[1]

Meanwhile, to facilitate the development of the "next generation" of tactical aircraft, the Air Force Flight Dynamics Laboratory initiated the Advanced Fighter Technology Integration (AFTI) program in 1973. AFTI was intended to demonstrate technologies that might be considered too high risk for inclusion at the full-scale development stage of a major weapon system acquisition.

> The intent of the Advanced Fighter Technology Integration (AFTI) program is to find a mechanism for the orderly transition of new technology into future aircraft.... Technologies that may be flight evaluated in USAF's advanced fighter technology integration

program would be applicable to future military and civil aircraft and might be retrofitted into some existing programs....[6]

The AFTI program was not specifically tied to the ATF program, but the Advanced Tactical Fighter was identified from the outset as one of the future aircraft likely to utilize AFTI technology.[6]

TARGET ACQUISITION AND WEAPON DELIVERY

Only one contracted ATF effort was conducted during 1974. This was the Target Acquisition and Weapon Delivery (TAWD) study. The 1972 ATF parametric and point design studies had considered several air-to-surface attack missions flown at supersonic speeds. Analysis showed that conducting an attack at supersonic speeds could substantially increase the survivability of the attacking aircraft, but the 1972 studies did not address issues of longer release ranges, reduced pilot reaction time, and target acquisition for a supersonic attack. TAWD was intended to address these issues. ASD released the request for proposal (RFP) for the TAWD study in October 1973, and seven companies responded. McDonnell Douglas was awarded a $200,000 contract in February 1974 and completed the study in October 1974. ASD briefed results to TAC in February 1975.[1]

PROPOSED FISCAL YEAR 1977 ADVANCED TACTICAL FIGHTER PROTOTYPE EFFORT

Based on the accumulated results of the TAWD and earlier studies, Headquarters Air Force Systems Command (HQ AFSC) developed a plan during early 1975 to prototype the ATF. The plan, involving $378 million over the period from fiscal year 1977 through fiscal year 1981, was to fund two contractors to build two prototype aircraft each, including "…airframe, engines, avionics, ECM, weapons system, etc., *for a complete integrated weapons system.*"[7] The prototypes, still envisioned as leading to a replacement for the F-4 and F-111, would include the following design features and capabilities:

1) Supersonic weapons delivery (requiring the use of improved weapons).
2) Advanced technology (1980 state of the art) engine.
3) All aspect ECM.
4) Integrated digital avionics for target acquisition and weapon delivery.
5) Advanced cockpit design for single crew member operation.

Adequate funding was not provided, and by mid-1975 the effort was abandoned, at least until more solid support could be generated for prototyping and for the ATF program in general.[1]

By this time, TAC had withdrawn ROC 301-73 for substantial revision, following an unsuccessful attempt late in 1974 to obtain funding for ATF development starting in fiscal year 1976. Major problems with the ROC included the following:

1) The threat assessment needed to be updated. For perspective, the October 1973 war in the Middle East had occurred since the ROC was originally written; this war provided a dramatic demonstration of the capabilities of Soviet-built air defense systems against the latest Western tactical aircraft.
2) Existing capabilities/deficiencies needed to be reassessed in view of the introduction of the Air Combat Fighter (ACF—the General Dynamics F-16 was selected for this role following the ACF competition that evaluated proposed production variants of the YF-16 and the Northrop YF-17) to the tactical force structure.
3) ACF would also adversely impact the availability of development funds for the ATF.

During 1975, TAC developed a revised version of the ROC and sought comments and input from HQ USAF, HQ AFSC, ASD, the Air Force Wright Aeronautical Laboratories (AFWAL*), and other appropriate Air Force organizations. However, it appears that this version also never completed the validation process.[1]

Since early 1974, the Air Force Flight Dynamics Laboratory (AFFDL, part of AFWAL) had sponsored studies by General Dynamics and McDonnell Douglas to help focus laboratory technology on an advanced air-to-ground fighter. The studies were called Air-to-Ground Advanced Fighter (ATGAF).

When direction came in 1975 to support the ATF ROC revision, the ATGAF studies were expanded and renamed Advanced Tactical

---

* Now part of the Air Force Research Laboratories.

Fighter Technology Evaluation and Integration. Funding was provided for a $2.1 million study program in 1976–1977, and an RFP was released in February 1976. Government participation grew to include ASD, the Armament Development Test Center (ADTC) and the Air Force Armament Laboratory (AFATL), NASA, and TAC.[1]

This may be the first study in which signature reduction began to be considered as an important part of the ATF concept. In early 1976, the following was reported:

> Air Force will initiate a series of conceptual design studies of an advanced tactical fighter (ATF) this spring as part of an overall service-wide effort to explore ways of accomplishing tactical interdiction missions....
>
> The Air Force plans to pick three separate contractors for parallel studies to get underway in late April or early May [1976]....
>
> Besides the overall design studies, the Flight Dynamics Laboratory intends to sponsor two other ATF studies in the area now generally called Stealth.... Flight Dynamics Laboratory had been asked to use Northrop and Lockheed as its Stealth contractors because these two companies have been working on a Stealth fighter aircraft program sponsored by Defense Advanced Research Projects Agency. But the laboratory is planning to compete the two ATF Stealth awards.
>
> The intent of the Stealth studies will be to explore how far it is practical to progress in the application of signature reduction techniques...without compromising the system's operational capability. They also will seek to identify how much of the remaining signature could be compensated by electronic countermeasures....
>
> The outcome of the year-long conceptual design studies is expected to be configurations, cost effectiveness data and projected payoffs to the ATF of new technologies. The latter include such items as standard avionics modules, active controls, composite materials and new cockpit ideas.... From data gathered during the studies USAF hopes to hammer out jointly with the contractors a set of detailed development plans for key technologies that need greater emphasis for application to the ATF.[8]

By the time the study contracts were actually awarded, the program was renamed Air-to-Surface (ATS) Technology Evaluation and Integration.[9] It is not clear whether separate "stealth" studies were ever contracted apart from the primary ATS design studies, as suggested above; however, the ATS design study contractors did specifically address "stealth." Survivability was clearly a central consideration in ATS.

Contracts of approximately $700,000 each were awarded to Boeing, Grumman, and McDonnell Douglas in June 1976. Three unsuccessful bidders—General Dynamics, Northrop, and Rockwell—also elected to participate on company or internal research and development (IR&D) funds. The studies addressed long and short range interdiction missions for an all-weather, manned, air-to-surface strike aircraft. Because of the continuing importance of weapon delivery issues, each airframe contractor worked with one or more weapons subcontractors. Various technology availability dates (TADs—considered to be when the technology is available for full-scale engineering development entry) were considered: Grumman evaluated technology available in 1981 and 1986; Boeing concentrated on late 1980s; and McDonnell Douglas studied a TAD of 1990. The companies developed candidate designs emphasizing various characteristics (e.g., supersonic cruise, short takeoff, high maneuverability, stealth) and identified critical technologies and required technology development activities. The funded ATS contractors delivered their final briefings in October 1977, and the unfunded participants presented their briefings in November.[1,2,10,11] The collective ATS results emphasized the following characteristics[5]:

1) High altitude, supersonic speed, and ECM will be needed to survive Mach number = 1.6 to 2.2; altitude = 50,000 to 65,000 ft; basing 250–400 n miles from targets, penetration up to 250 n miles required; and SAMs are the primary threat.
2) New air-to-surface weapons are required to develop the full potential of advanced aircraft.
3) Stealth is a "promising technology."
4) Internal/conformal weapon carriage needed (for stealth and fast penetration).
5) Composite materials reduce cost and weight.
6) Advanced avionics are needed for adverse weather operation [synthetic aperture radar (SAR), moving target indication (MTI)].
7) Integrated system display is needed for one-man operation.

Many of the attributes identified by this time would prove to be equally applicable to an advanced air superiority fighter because they were based on the need to survive in the future threat environment. ATS correctly forecast that the capability to survive and fight effectively in the late 1980s and beyond would only be achieved through the application and integration of a spectrum of advanced technologies. To this end, an important contribution of the ATS studies was that they helped to establish ASD/AFWAL cooperation, which would be crucial to subsequent ATF technology development.[1]

OFFENSIVE AIR SUPPORT MISSION ANALYSIS

At essentially the same time as the ATS studies were conducted, the ASD Deputy for Development Planning (ASD/XR) conducted a series of air-to-surface mission analysis activities. The study began in 1975 as the Close Air Support Mission Analysis. Night and adverse weather considerations were an important part of this study. During 1976, the study was broadened to include the Battlefield Interdiction mission and was renamed Close Air Support/Battlefield Interdiction (CAS/BI) Mission Analysis. In October 1976 the scope was expanded again, and the name was changed to Offensive Air Support Mission Analysis (OASMA). Offensive air support (OAS) is a term used within the North Atlantic Treaty Organization (NATO); it includes close air support, air interdiction that directly impacts the ongoing land battle, and certain tactical air reconnaissance missions. The study was completed and results were briefed to Air Force leadership during 1977.[1]

However, the study did not address air-to-surface counterair missions (such as missions conducted against enemy airfields) or deep interdiction, which was envisioned as an important mission for the ATF. Furthermore, the study focused on potential near-term subsystems improvements or other modifications to existing aircraft for the period 1977–1990. For example, two-seat variants of the F-16 and A-10 (seen in Fig. 4) with enhanced avionics and sensors were identified as approaches to providing an all-weather capability for the missions that were studied. The study therefore had limited impact on the ATF program, other than to provide, as a background, an improved understanding of what could be achieved with modifications to existing aircraft.[1]

STRIKE SYSTEM STUDY

The Strike System Study (S[3]) was initiated as a joint TAC/AFSC effort to lay the groundwork for the development of one or more "next-generation" strike weapon systems. The focus was entirely air-to-surface and was relatively far-term [i.e., initial operational capability (IOC) 1990 or later]. Planning for this study began in December 1976 and continued through 1977. S[3] was funded under program element 63230F, "Combat Aircraft Technology" (this is the program element that would eventually fund the Advanced Tactical Fighter program). The first program management directive (PMD) for this program element was issued in January 1977, a revision in June 1977 gave recognition (and allocated funding) to the S[3] effort.

S[3] was one of the first Air Force programs initiated under Office of Management and Budget (OMB) Circular A-109, which was released

Fig. 4   Two-seat variants of the multirole capable F-16 fighter and the A-10 close air support aircraft, equipped with improved avionics, were identified in the OASMA study as approaches to provide enhanced night/all-weather attack capability.

on April 5, 1976, and implemented by the revised DODD 5000.1 on January 18, 1977. Under these new acquisition regulations, a new major program would only be authorized following the approval of a mission element need statement (MENS) by the secretary of defense.* Accordingly, the preparation of a MENS was specifically an objective of S^3.

Initially the S^3 study was very broad, considering vertical/short takeoff and landing (V/STOL) aircraft, unmanned aerial vehicles (UAVs), and "system-of-systems" approaches in addition to conventional aircraft solutions. To reflect this broad emphasis, the name was changed to

---

* A MENS is "a statement prepared by HQ USAF to identify and support the need for a new or improved mission capability.... The MENS is submitted to the SECDEF [Secretary of Defense] or SAF [Secretary of the Air Force] for a Milestone Zero decision" (AFR 57-1, June 12, 1979). The intent of a MENS was to focus on the needed capability, rather than on any particular solution or system.

Synergistic Strike System (still abbreviated as S³) in late 1977. In October 1977, the Directorate of Mission Analysis (XRM) was activated under ASD/XR, the deputy for Development Planning, and assumed responsibility for the S³ effort.

However, there was little progress in producing a satisfactory MENS for such a broad effort. By February 1978 the fourth draft of the S³ MENS was in coordination. During early 1978 the scope was narrowed based on a perception that the Office of the Secretary of Defense (OSD) would only approve a more focused MENS. At a joint TAC/ASD/Air Staff meeting in June 1978, it was decided that the Air Staff would rewrite the MENS, "focusing more tightly on night, all-weather capability against mobile battlefield interdiction targets, hopefully weakening the argument for unmanned systems."[1]

In July 1978, S³ was absorbed into a two-part effort to address both near-term and far-term improvements in tactical strike capabilities. The overall effort was called improved tactical attack system (ITAS), and encompassed a new Enhanced Tactical Fighter (ETF) component to meet the near-term (late 1980s) need, and the advanced tactical attack system (ATAS), which was essentially a continuation of the S³ effort to meet the far-term need. The ITAS mission analysis activities were managed, as S³ had been, by ASD/XRM, the Directorate of Mission Analysis.[1]

From mid-1978 through late 1979, the ETF and ATAS efforts made little progress, other than program planning and definition. A contract for technical analytical study support (TASS) for ATAS was awarded to the BDM Corporation in December 1979, but there were still uncertainties regarding program direction and funding.[1]

INTRODUCTION OF AIR SUPERIORITY AS A POTENTIAL ADVANCED TACTICAL FIGHTER MISSION

THREAT DEVELOPMENTS

By that time a specific air-to-air threat had begun to coalesce in the form of a new generation of Soviet fighters. Test aircraft referred to as the "Ram-K" and "Ram-L" were sighted at the Ramenskoye test facility at Zhukovskiy (near Moscow) in 1977 and 1979. These were obviously prototypes of advanced fighters, later identified as the Su-27 and MiG-29, which were considered comparable to the F-15 and the F/A 18, respectively. These threatened the technological superiority on which the United States depended to offset the Warsaw Pact's superior numbers. Also in 1978, the Soviets claimed to have conducted a suc-

cessful test of "look-down, shoot-down" radar on an advanced derivative of the MiG-25 interceptor (which later emerged as the MiG-31),
achieving a theoretical kill from an altitude of 6000 m (approx. 20,000
ft) of a target flying at 60 m (200 ft). This capability would be a serious
threat to low-flying NATO interdiction aircraft and bombers.[12] These
emerging Soviet threats are depicted in Fig. 5.

When ATF-related activities had begun around 1970, the F-15 was
expected to be able to meet the air-to-air threat for the foreseeable future. However, by 1978 there was evidence that Soviet fighters would
soon be able to match the F-15, and so there was a clear need on the
horizon for a "next-generation" U.S. Air Force air superiority fighter to
enter service in the mid-1990s.

RESTRUCTURING OF THE ADVANCED TACTICAL FIGHTER PROGRAM

As a result, planning began in late 1979 for a restructured ATF program that would consider air-to-air missions in parallel with air-to-
surface missions. The effort would encompass[13] 1) design and technology studies, 2) mission analysis—air-to-air and air-to-surface in
parallel—and 3) development and validation of mission element need
statements (MENS)—air-to-air and air-to-surface in parallel—to support an anticipated Milestone Zero decision.

The new direction was formalized in April 1980 with a new version
of the program management directive (PMD) for combat aircraft technology. PMD R-Q 7036(4)/63230F directed several important changes.
First, it eliminated the near-term ETF effort. Second, it changed the
name of the ATAS project to Advanced Tactical Fighter (ATF) (the
overall heading was still called combat aircraft technology). The new
PMD eliminated consideration of unmanned systems but added the
air-to-air role. The scope of the ATF program starting in April 1980 was
therefore to study manned tactical aircraft systems for air-to-surface,
air-to-air, or both. The reasoning behind the program direction at that
time is described as follows[13]:

> The purpose of the Advanced Tactical Fighter Program (ATF) is
> to focus on development of concepts and proving technology for
> the next generation tactical fighter aircraft. The ATF program fo-
> cuses on the capabilities required in manned weapons systems
> which will permit the U.S. Tactical Air Forces to carry out their mis-
> sions well into the next century....
>
> United States's vital interests are worldwide and so are potential
> TACAIR commitments. The most demanding environment for
> our forces is expected in a NATO–Warsaw Pact war in Central

Fig. 5   By the late 1970s, it was recognized that advanced Soviet-developed fighters would be fielded by the Soviet Union in the 1980s in the form of the MiG-29 (top), the Su-27 (middle), and the MiG-31 (bottom), and that these aircraft would threaten theater air superiority.

Europe.... It is a war we cannot afford to lose, but other scenarios are becoming increasingly important and must also be considered when we address the capabilities required in our next generation fighters.

The length of time associated with defining and proving concepts and developing a new generation aircraft is also a major consideration in shaping a fighter development program. With current R&D management practices, a new aircraft would not enter the inventory in significant numbers before the mid-1990s and most of its service life will be spent in the twenty-first century....

THE REQUIREMENT: For the past several years tactical requirements have been developed by mission area. This view has been helpful in dividing the spectrum of tactical requirements into manageable areas for programming and budgeting. It also directly relates required functions and tasks to assigned missions.

The mission area structure has been less helpful in efforts to forecast and clearly articulate tactical aircraft needs. The budget process...often requires that very disparate systems and concepts be compared. It is relatively easy to compare alternative single-purpose systems. Fighters and TACAIR, on the other hand, are general purpose systems and...affect more than one mission area and their relative contributions and cost trades become even more difficult when considered only within a single mission area. In an effort to gain a clearer picture of the total tactical aircraft contribution, we are asking the question, "What is the requirement for new tactical aircraft" as well as, "What is the individual requirement in each mission area?" This broader perspective has helped articulate the real world requirements for future tactical fighters. Specifically, it has allowed us more flexibility to consider force numbers, force structure, uncertainties in predicting the threat, and technological opportunities to lower costs while meeting well defined performance objectives. This duality of thinking [looking at the need across all mission areas, together with individual mission area needs] and program management is reflected in the two ATF Mission Element Need Statements (MENS), one for air superiority mission requirements and another for strike fighter requirements. The MENS are clearly tied together and a single Program Management Directive (PMD) has been issued for the program. The PMD requires parallel consideration of the two mission area requirements and that conceptual solutions include multi-mission aircraft as well as single-purpose solutions, at least through M/S I [Milestone I, or Dem/Val start].[13]

Air-to-surface was noted to be the most pressing need if one were to look at the *immediate* outlook in each mission area. The recent development of the F-15 and the multirole F-16 met the near-term need in the air superiority mission area. However, a shortfall of U.S. capabilities in the air superiority mission area could reasonably be expected before the turn of the century due to advances in the Soviet threat. Consideration of trends in the acquisition cycle—then recognized to be approaching 12 to 14 years—together with budget constraints, suggested that the Air Force might only be able to develop and field one tactical fighter aircraft by that time. Premature commitment to meeting the near-term air-to-surface mission area need could therefore leave a deficiency in the air superiority area. Failure to achieve air superiority would, in turn, result in an inability to perform all other tactical air missions. Accordingly it was decided that, for the time being, both options should be kept open:

> The Advanced Tactical Fighter (ATF) program is structured to bring air-to-surface and air-to-air aircraft options up to, but short of, engineering development. When threat driven development decisions are required, the relatively low cost but time consuming (pre-MS II) front end of the acquisition cycle will be complete. This approach...avoids premature commitment to solving one mission area problem to the exclusion of another.
>
> The program is clearly threat and cost driven. Although uncertainty exists about the future threat, we must be technologically ready when a threat-driven development decision is required. Also, we must be fully aware of the cost tradeoffs and we must know what must be given up to purchase each additional increment of capability. The program is structured to answer these questions.[13]

Initial schedule milestones were full-scale development (FSD) start in 1987, and IOC in 1993 or 1994.[14] Regardless of whether the ATF turned out to be a strike aircraft or an air superiority fighter, there were certain attributes it was likely to have. Both would draw from the same set of emerging technologies. Both would have to operate from the same set of (possibly damaged) bases into a very high threat environment. It was noted that "survivability considerations are likely to be dominant" for any future tactical aircraft.[13]

Affordability and supportability were also important considerations from the outset:

> "We want to maintain a competition between the force availability and the system capability in the advanced tactical fighter program," one USAF program official said. "The force availability is a function of acquisition costs, maintenance, reliability and reparability. System capability, on the other hand, is based on the

radar signature, vulnerability, performance and weapons systems."[14]

Perhaps the most important contribution technology can offer
the next generation fighter is significant improvements in affordability, maintainability, availability, and reliability. Lower operating
costs, while meeting required performance objectives, are very important considering current fiscal realities. Maintainability, availability, and reliability are important because, during wartime, engaged forces are what count—not forces on the airfield, but forces
in the air.[13]

These statements reflect an understanding of the value of supportability, both to reduce costs in peacetime and also to improve effectiveness
in wartime.

Although the primary threat to U.S. security at this time was obviously the Warsaw Pact forces in Europe, the ATF concept was shaped
by other considerations as well. In particular, the ATF was expected to
have a long range to operate effectively in theaters that are more geographically dispersed than Central Europe (such as the Middle East).

TACTICAL FIGHTER TECHNOLOGY ALTERNATIVES AND 1995 FIGHTER STUDY

Two Air Force Flight Dynamics Laboratory studies begun in late
1979 reflected the new parallel emphasis on air-to-air and air-to-surface
missions. The Tactical Fighter Technology Alternatives (TAFTA) study
was essentially a follow-on to the ATS study described previously and
concentrated on the air-to-surface mission. TAFTA placed additional
emphasis on compatibility with existing as well as advanced weapons
and on reducing the dependence on fixed bases with long runways.
Boeing and Grumman each developed advanced strike aircraft concepts under 18-month contracts that began in September 1979.[1,14,15]

The 1995 Fighter Study ran in parallel with TAFTA and concentrated on the air-to-air missions that were expected for that time period. This may have been initiated as a result of the recent developments in the air-to-air threat already discussed. In the 1995 Fighter
Study, General Dynamics and McDonnell Douglas developed air-to-air
advanced fighter concepts.[14,15]

TAFTA and the 1995 Fighter Study were managed by the Air Force
Flight Dynamics Laboratory. The approach to both of these studies was
similar to the ATS studies in that the contractors examined "extremes
in aircraft concepts that range from high- to low-altitude penetration at
supersonic cruise speeds, to Stealth technology" and also included designs that were optimized for short or vertical takeoff and landing.

Both TAFTA and the 1995 Fighter Study were completed in early to mid-1981.[14,15]

By that time several other events were in progress, which would culminate in a Milestone Zero decision and the start of the formal concept exploration phase of the ATF program.

PREPARATION FOR MILESTONE ZERO

MISSION ANALYSIS

Under the revised ATF program, the ongoing ATAS mission analysis (ATASMA) continued and was joined by a parallel air-to-air effort, the Advanced Counterair Engagement Mission Analysis (ACEMA).[1] "The pre-Milestone Zero mission analysis is structured to answer the sometimes deceptively simple questions required to support the mission element need statement."[13] These questions included the following:

1) What kind of fighter should be developed?
2) What will the dominant need be in 1995?
3) What can technology support?
4) What is affordable?

The mission analysis encompassed the following topics[13]: 1) scenario development [Europe, Southwest Asia, Korea, and Continental United States (CONUS)/Worldwide]; 2) mission needs and essential tasks; 3) threat projection and assessment; 4) technology assessment; 5) generic aircraft design packages; and 6) parametric trade studies.

Both ATASMA and ACEMA used previous TAC mission area analyses as a point of departure and were intended to lead into industry concept definition and design studies as well as to support the development of the mission element need statements (MENS).[1,5]

ADVANCED TACTICAL ATTACK SYSTEM MISSION ANALYSIS. The overall aims of ATASMA were to assess tactical air power (TACAIR) air-to-surface deficiencies and to identify and evaluate the potential of technologies, operating concepts, and generic alternatives to meet these deficiencies. "Generic alternatives" were defined to include additional quantities of current weapon systems, modification of existing systems with new technology, as well as generic designs for new weapon systems. The overall framework of the study was the need for U.S. tactical aircraft to defeat a Warsaw Pact breakthrough in Central Europe; however, "a sensitivity to operational employment is being addressed to identify problems and deficiencies unique to the other theaters and their respective environments." The other theaters addressed in this study were the Persian Gulf and Korea.[5]

Key issues considered in the ATASMA were[5] 1) aircraft survivability (encompassing penetration speed/altitude, observables, and ECM), 2) weapon lethality (in particular night/adverse weather targeting capability), and 3) sortie generation (including takeoff/landing field length and supportability for austere/damaged base operations).

The need for short takeoff and landing (STOL) capability was expressed at this time (1980) to allow the ATF to operate from bases that were under attack. It was noted that with "current repair capability" of 4 craters every 3 h (Ref. 5) and current field length requirements of 3000–4000 ft, one attack per day on runways would reduce the sortie generation rate (SGR) of U.S. TACAIR forces almost to zero. On the other hand, if the required field length could be reduced to 2000 ft, this would allow the SGR to be about 60% of its maximum unimpaired value. This was about the same SGR that would result if the enemy attacks were directed against aircraft shelters or maintenance facilities, rather than runways. Thus there would be no single weak link.[5]

European weather was seen as another factor that would have a significant effect on sortie generation. "A Soviet attack timed to occur during a period when meteorological conditions could disrupt air terminal operations would degrade or negate U.S. operations against the attack."[5] Because NATO doctrine relied heavily on airpower to negate the Warsaw Pact's numerical advantage in ground forces, this was a serious issue. The ability to conduct flight operations, as well as to accomplish target acquisition and weapon delivery, at night or in adverse weather would be critical.

ATASMA also included an effectiveness analysis of "quality vs quantity." The study findings were firmly in favor of quality. Problems associated with inexpensive "numbers" fighters, which had been advocated for over a decade by a group known as the "lightweight fighter Mafia" within DOD and the Air Force, included the following:

1) limited range—inability to reach certain targets at all would negate the advantage of numbers;
2) small weapons load;
3) high rate of attrition in European combat environment, including pilot attrition; and
4) logistical problems with large numbers of aircraft, flight crew, etc.

The bottom line on this issue was that only a highly capable aircraft could perform certain critical mission tasks *at all,* and numbers would never make up for key capability shortfalls.

ADVANCED COUNTERAIR ENGAGEMENT MISSION ANALYSIS. An air-to-air strategic defense mission analysis activity was already under way in ASD/XRM as of late 1979. As a result of the April 1980 Combat Aircraft Technology (CAT) PMD revision, the strategic defense mission analysis was expanded to include all air-to-air missions; the name was changed to Advanced Counterair Engagement Mission Analysis (ACEMA), and it became the "sister study" to ATASMA within the overall ATF effort.

A TASS contract for the ACEMA study was awarded to General Research Corporation (GRC) by mid-1980. From May to September, ACEMA examined the cost-effectiveness of various aircraft–missile combinations both in Central European and North American (air defense) scenarios. An important finding was that "for North American air defense the study results noted a major gain in effectiveness from the combination of supercruise capability with a good autonomous intercept capability."[1]

Subsequently ACEMA also addressed the Southwest Asian (Persian Gulf) and Korean theaters. In the various scenarios a total of eight aircraft were evaluated including three existing aircraft or variants and five new designs. The five new aircraft included one lightweight (30,000 lb takeoff gross weight) fighter, two medium weight (approx. 50,000 lb) designs, and two very large (up to 100,000 lb) fighters. The designs included characteristics such as STOL, high maneuverability, large weapons load, and/or long loiter performance. All five new concepts included a supersonic cruise capability.[1]

In October 1980 Congressional action eliminated all ATF funding for FY81. Air Staff provided $700,000 in November 1980 to continue the ATASMA and ACEMA efforts, although this was insufficient to accomplish all planned activities. Reprogramming of an additional $650,000 in April 1981 allowed the ATASMA and ACEMA efforts to be completed by October 1981.[1]

FUTURE FIGHTER ALTERNATIVES STUDY

In August 1980, a Future Fighter Alternatives Study (FFAS) was initiated by HQ AFSC. At its initiation, this study was not specifically related to ATF, but rather was intended to support the preparation of the FY83 Air Force Program Objectives Memorandum

(POM).* FFAS attempted to address near-term and mid-term capability improvements, as well as new fighter concepts and development paths for the far term. The intent was to take a relatively unconstrained look at what would be needed 5, 10, 15, and 20 years in the future to accomplish the "canonical Air Force tasks": gain/maintain air superiority, interdict air/sea/ground lines of communication, and blunt enemy forces arrayed against friendly surface forces.[16]

A fairly elaborate methodology was laid out initially, involving a "mission volume" that represented the product of the number of aircraft, the mission availability, and the mission capability for each aircraft type considered. The total "mission volume per dollar" would provide the basis for ranking the various aircraft. As a *starting point,* 17 aircraft types (existing and derivative) were to be considered for near-term and mid-term derivatives. New aircraft concepts would also be looked at for the far term.[17]

FFAS appears to have been heavily controlled by HQ AFSC, although there was participation from other agencies and commands. TAC participants expressed serious reservations about the study scope and methodology, noting that the focus was too broad; too many "non-player" aircraft were being considered as mid-term derivatives; force structure realities and affordability were not being properly addressed; and the methodology appeared biased in favor of multimission capable aircraft.[17]

It is not clear whether all of the analysis was carried out as planned. At a FFAS General Officer Steering Committee meeting on October 30, 1980, the near-term portion of FFAS was narrowed down to concentrate on potential improvements to the F-15 and F-16, *specifically for the air-to-surface mission.* The far-term portion of the study, to be completed in early 1981, would involve a "more generalized review of the total fighter force through the 1990s."[18]

Adding to, or possibly obviating the need for, the far-term portion of FFAS were the following efforts that reached completion during the period of the FFAS study:

1) ASD/XRM managed the ACEMA and ATASMA parallel mission analyses for air-to-surface and air-to-air roles, respectively.
2) AFFDL managed TAFTA and the 1995 Fighter Study.

---

* The POM is defined as the "service request for resources to accomplish its mission." Each service submits a POM to the Office of the Secretary of Defense (OSD) as part of the two-year budgeting cycle.

3) Air Staff developed and coordinated a MENS for a new fighter aircraft.

On October 23, 1981, approximately one year after the study got under way, FFAS ended with the conclusion that "enough pre-Milestone Zero studies have been conducted, and it is time to direct efforts toward concept exploration phase of the ATF program."[19] This echoed the sentiment expressed at an Air Force Systems Command Council review of the ATF program on the previous day. It was noted that the MENS had already been submitted to OSD, approval of which would constitute Milestone Zero [as will be described]. Furthermore,

> The [FFAS] Steering Committee specifically endorsed an ATF acquisition strategy that supports focused technology thrusts (as opposed to near-term competitive prototyping) to preserve STOL, low observables, and efficient supersonic cruise options for ATF designers.[19]

The issue of prototyping would arise repeatedly during the next few years, but the prevailing view going into concept exploration was that a period of focused technology development would be needed in order to achieve the capabilities envisioned for the ATF and that near-term competitive prototyping would "freeze the technology" too early.

MISSION ELEMENT NEED STATEMENT AND MILESTONE ZERO

In conjunction with ATASMA and ACEMA, and in preparation for the anticipated Milestone Zero decision, an Air Staff "MENS Working Group" issued two draft ATF mission element need statements in July 1980, one for air-to-surface ("Advanced Tactical Strike Fighter") and one for air-to-air ("Advanced Air Superiority Fighter"). The plan was to carry the two sets of requirements in parallel at least through the completion of "Phase 0," concept exploration.

Under the recent acquisition guidelines (OMB Circular A-109, etc.), the MENS was the primary document justifying the need for a new capability. In the Air Force, the Air Staff was responsible for preparation of the MENS for a new weapon system. A "For Comment Draft MENS" would be circulated within the Air Staff and to the using command (in this case TAC), Air Force Logistics Command and AFSC. Based on the comments received, a "For Coordination MENS" would be prepared, coordinated, and submitted to the secretary of the Air Force (SAF) for approval (of programs within SAF approval authority) or for recommendation to SECDEF (in the case of major programs such as ATF). Approval of the MENS, at whatever level was required, would constitute Milestone Zero and the beginning of the formal concept exploration phase of a program.[20] A program management directive (PMD) would then be issued, providing formal guidance for the new program.

Accordingly, the two draft ATF MENSs were circulated for review during 1980 and early 1981. In May 1981, a single *Advanced Manned Tactical Aircraft* (AMTA) MENS was proposed that covered both the air-to-air and air-to-surface mission areas.[21] Elements from the AMTA MENS were incorporated into a new *MENS for New Fighter Aircraft,* which was issued on July 6, 1981, for Air Staff coordination and submission to OSD. Key characteristics stated in the *MENS for New Fighter Aircraft* are summarized here:

1) *Mission Areas:* strategic air defense, counter air, and close air support/interdiction.
2) *Mission Element Need:*
   To meet the air-to-air and air-to-surface fighter missions requirements during the 1990–2010 period, the new fighter aircraft must have performance balanced with logistic supportability and be capable of
   a) Destroying hostile aircraft during all environmental conditions before they can attack friendly air or surface targets
   b) Seeking, detecting, tracking, identifying, engaging and destroying hostile ground targets during all environmental conditions
   c) Worldwide response
   d) Austere basing including damaged runways/less than full support
   e) Enhanced survivability "by capitalizing on emerging technologies," i.e., low observables
   f) Tracking and engaging multiple targets simultaneously
   g) High mission capable rates with reduced manpower and spares

On September 1, 1981, James E. Williams, the acting assistant secretary of the Air Force (Research, Development and Logistics), sent the *New Fighter Aircraft* MENS to the Office of the Undersecretary of Defense (Research and Engineering) (OUSDR&E) for approval. As noted previously, the AFSC Council and the FFAS General Officer Steering Committee met on October 22 and 23, respectively, and expressed their views that enough pre-Milestone Zero studies had been completed and it was time to move on with the concept exploration phase of the ATF program. On November 18, the proposed ATF program was briefed at OSD. The MENS was discussed at this meeting. Then, on November 23, the Defense Resources Board (DRB) met to review major system new starts. One of the new starts approved at this meeting was the ATF. This event is considered to be Milestone Zero for the ATF program.

Following the November 18 briefing, the *MENS for New Fighter Aircraft* was circulated for comment to appropriate organizations within OSD. On January 11, 1982, the comments were sent back to the assistant secretary of the Air Force (Research, Development and Logistics). The chief comments were

1) Very ambitious requirements. MENS implies that one system can meet all of them. "The possibility that more than one design may be required needs to be addressed."[22]
2) Additional study is required to determine which missions are most critical and which specific set of performance characteristics will be needed.
3) The MENS is too general—recommend breaking it up [again!] to cover the different mission areas explicitly.
4) Needed capabilities include many characteristics (STOL, performance, cost, range, supportability) that have traditionally had to be traded off against each other; need to address how this will be dealt with in the ATF program.

Various additional specific comments and recommended changes were stated. Nevertheless, the bottom line was, "The need, as expressed in the MENS, is supported and appears justified."[22]

However, in November 1981, House–Senate conferees denied fiscal year 1982 funding to ATF. Thus, the status at the end of 1981 was that the ATF program was approved but unfunded.

ENGINE STUDIES

During this time, studies were conducted in parallel to evaluate possible propulsion systems for future Air Force and Navy aircraft. These studies were essential to defining the technologies as well as the size and cycle of the ATF engines.

The Advanced Technology Engine Studies (ATES) was a Navy sponsored program that ran from 1980 to 1982 and was not tied to a specific weapon system. It evaluated propulsion systems for a wide range of

possible future Navy and Air Force aircraft, including advanced
fighters for both services. After ATES, the ASD Aero Propulsion
Laboratory (APL) funded the Propulsion Assessment for Tactical
Systems (PATS) studies, which ran from September 1982 to September 1983. This study focused more narrowly on Air Force and Navy
fighter aircraft and helped define the initial engine concepts for
the ATF.

REQUEST FOR INFORMATION

One final activity was initiated prior to Milestone Zero, which would
help bridge the gap between program approval and program funding.
This was the Advanced Tactical Fighter Request for Information (RFI).

In October 1980, an ad hoc committee of the ASD Advisory Group
(a subset of the Air Force Scientific Advisory Board) had met to review
the scope and capabilities of ASD/XRM and determine how the effectiveness of the relatively new XRM organization could be maximized
(XRM was established in October 1977).[23] The committee received
briefings on the organization as a whole, as well as on the specific mission analysis activities that were under way, including the ATASMA and
ACEMA studies mentioned previously. One of the Advisory Group's
recommendations was to "take advantage of the 'inherent clout' of ASD
getting industry involved in [XRM's] activities at no cost."[23] Almost simultaneously with the Advisory Group's meeting, Congress had denied
fiscal year 1981 funding for the ATF program, so the recommendation was extremely timely. It was determined that an RFI would be the
best way to obtain such no-cost industry participation in the ATF program.[1]

Accordingly, on May 21, 1981, ASD/XRM issued an RFI for the ATF
to nine companies: Boeing, Fairchild, General Dynamics, Grumman,
Lockheed, McDonnell Douglas, Northrop, Rockwell International, and
Vought. Industry participants were asked to submit conceptual designs
and to prepare position papers on a variety of technical and operational issues that would be identified by the government. Technology
content was to be consistent with an IOC of early to mid-1990s. Cost estimates were to be based on possible production runs of 120, 500, or
1000 aircraft.[1]

APL released an ATF engine RFI the following month. The engine
RFI stated the desired attributes: supersonic persistence without afterburner (supercruise), STOL distances of 1500 ft, stealth, a reduced cost
of ownership, and a targeted system IOC of 1993.

The RFI process was initially expected to take nine months, with mid-term and final reports to be provided by the participants. The government would then provide feedback to the industry participants within 60 days, collectively and/or individually depending on the extent of proprietary information involved. The RFI was seen as contributing to the basis for a Milestone Zero decision, and as such would prepare for, not take the place of, the formal concept exploration effort (Phase 0).[24,25]

For perspective, the parallel for-comment Advanced Air Superiority Fighter (AASF) and Advanced Tactical Strike Fighter (ATSF) MENSs were in circulation and were considered to be applicable at the time of RFI release. Very shortly afterward, however, the requirements were merged into a single New Fighter Aircraft (NFA) MENS that covered both the air-to-air and air-to-surface ATF missions. By the time RFI interim reports were presented (as described below), the NFA MENS had been validated by Air Staff and was in review at OSD.

At the RFI kickoff meeting on June 17, 1981, the industry participants were provided with classified threat scenarios, mission definitions, a technology database, and a more detailed definition of the information that was being requested. Nine specific issues were stated at this time for consideration and response by the participants:

1) aircraft/weapons performance trades (in particular addressing the impact of weapon standoff range on total system effectiveness);
2) quality vs quantity;
3) delaying vs destroying enemy second echelon forces;
4) one vs two crew members;
5) base level maintenance (reducing/eliminating intermediate-level maintenance, battle damage repair considerations);
6) observables/countermeasures tradeoffs;
7) specialized vs multirole design;
8) air base attack vs air-to-air kills; and
9) impact of concepts on propulsion system requirements.

The RFI gave equal emphasis to the air-to-air and air-to-surface mission areas. Options for accomplishing the two missions ranged from a single, multirole aircraft to a single aircraft with modular components (the weapons systems, or even the wings) to two completely different aircraft. There was, at the time, high level OSD interest in exploring the issues of specialization vs multirole capability in tactical aircraft; for example, in October 1981, it was reported that

> Top aides of Richard D. DeLauer, Under Secretary of Defense for Research and Engineering, are considering the cost advantages of increasing procurement of specialized tactical aircraft to reverse an existing trend toward procurement of multirole capability....
>
> Whether the Air Force should combine the attack and air-superiority functions of its advanced tactical fighter in one airframe "is a decision that must be made in the context of force structure at that time," Transue [Director of Air Warfare for Tactical Programs under DeLauer] said. The Air Force has drafted two mission element need statements and has funded studies by Grumman and Boeing of the attack role, and by McDonnell Douglas and General Dynamics of the air-superiority mission [referring to the "1995 Fighter Study" mentioned earlier].[26]

ASD/XRM was responsible for overall management of the RFI process. Air Force Wright Aeronautical Laboratories (AFWAL) participation was solicited at the time of RFI release to support the consideration of appropriate advanced technologies. This participation was formalized in a memorandum of agreement between AFWAL and ASD/XR, signed in September 1981. The Armament Division of AFSC (located at Eglin Air Force Base, Florida) was added via an amended version of the same memorandum, signed in January 1982. Other organizations playing significant roles in the RFI process included the Foreign Technology Division, the Air Force Acquisition Logistics Division, and Tactical Air Command, as well as other directorates within ASD/XR. For example, ASD/XRH, the Directorate of Design Analysis, reviewed the conceptual aircraft designs submitted by the contractors.[1,27] In conjunction with the RFI, ASD/XRM also conducted an in-house Advanced Tactical Fighter mission analysis (ATFMA), which was a follow-on to the earlier ATASMA and ACEMA studies. One of the things that ATFMA addressed, which the earlier mission analyses did not, was multirole considerations.[1]

The mid-term RFI briefings by industry were presented from November 16–19, 1981. As already noted, Milestone Zero approval was obtained on November 23, 1981, formally moving the ATF program into "Phase 0," or concept exploration. Results of the RFI are therefore described in the following chapter. A timeline of the period through (and slightly beyond) the RFI is shown in Fig. 6.

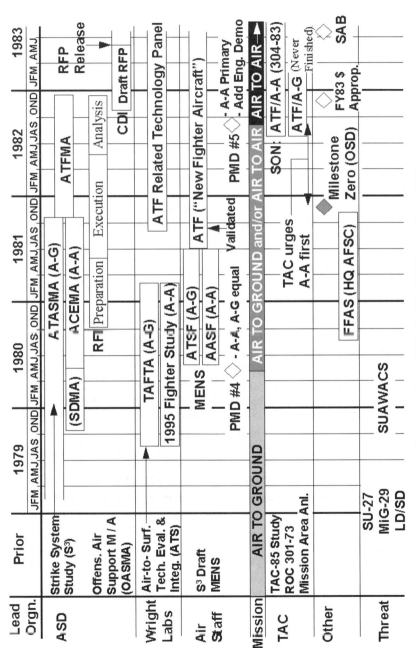

Fig. 6 Timeline—Transition from 1970s (A-G) ATF into modern ATF program.

REFERENCES

[1]Ferguson, P. C., *From Advanced Tactical Fighter (ATF) to F-22, Part I—To Milestone 0 and Beyond: 1970–1982,* Office of History, Aeronautical Systems Center, Wright–Patterson AFB, OH, May 1996.

[2]*Air-to-Surface (ATS) Technology Evaluation and Integration Study,* Grumman Aerospace Corp., Rept. No. AFFDL-TR-77-131, Bethpage, NY, Jan. 15, 1978.

[3]*Concept of Operation for the Advanced Tactical Fighter (ATF),* United States Air Force Tactical Air Command, Deputy Chief of Staff (Plans), Jan. 1971. SECRET. (Unclassified information only used from this source.)

[4]*Semi-Annual History,* AFSC Aeronautical Systems Div., Directorate of General Purpose and Airlift Systems Planning (ASD/XRL), Jan.–June 1972, p. 15.

[5]Sudheimer, R. H., (ASD), "Advanced Tactical Attack System Mission Analysis (ATASMA)," *Tactical Aircraft Research and Technology,* NASA-CP-2164, March 1981. SECRET. (Unclassified information only used from this source.)

[6]Miller, B., "USAF Plans Advanced Fighter Program," *Aviation Week and Space Technology,* July 16, 1973, pp. 14–16.

[7]Draft Memorandum, HQ AFSC/XRLA, Subject: *ATF Prototype,* April 21, 1975 (quoted in Ferguson, *From Advanced Tactical Fighter*).

[8]Miller, B., "Advanced Tactical Fighter Studies Set," *Aviation Week and Space Technology,* Jan. 19, 1976, pp. 16, 17.

[9]"Lockheed's Kelly Johnson Building 'Stealth' Aircraft," *Aerospace Daily,* Vol. 80, No. 16, July 23, 1976, pp. 121, 122.

[10]*Air-to-Surface (ATS) Technology Evaluation and Integration Study,* McDonnell Douglas Corp., Rept. No. MDCA4285-2, St. Louis, MO, Jan. 27, 1977.

[11]*Air-to-Surface Technology Integration and Evaluation Study,* General Dynamics, Rept. No. FZM-6667, Forth Worth, TX, May 11, 1977.

[12]Green, W., *New Observer's Book of Aircraft,* Frederick Warne and Co., 1985.

[13]Henry, D. D., Lt. Col. USAF (HQ USAF/RD&A), "Advanced Tactical Fighter Program," *Tactical Aircraft Research and Technology,* NASA-CP-2164, March 1981. SECRET. (Unclassified information only used from this source.)

[14]Robinson, C. A., "Air Force to Develop New Fighter," *Aviation Week and Space Technology,* March 30, 1981, pp. 16–19.

[15]Patterson, R., Capt. USAF, *Wright Laboratory's Role/History in the Advanced Tactical Fighter Technology Development,* Wright Lab., Rept. No. WRDC-TM-90-603-TXT, Wright–Patterson AFB, OH, Jan. 1990.

[16]Message from AFSC/CV to ASD/CC, Subject: *Fighter Alternatives Study,* Aug. 14, 1980.

[17]Wallace, Lt. Col. (TAC/DRFG), to TAC/CV, Staff Summary Sheet, Subject: *Future Fighter Alternatives Study,* Sept. 18, 1980.

[18]Adams, J., Col., to TAC/CC, /CV, /CS, /XP, Staff Summary Sheet, Subject: *Future Fighter Alternatives Study Steering Committee Meeting,* Nov. 3, 1980.

[19]Eaglet, R., Col., to HQ AFSC, HQ USAF, and others, Subject: *Future Fighter Alternatives Study,* Nov. 3, 1981.

[20]Air Force Regulation AFR 57-1, *Statement of Operational Need (SON),* HQ USAF, June 12, 1979.

[21]Smith, P., Brig. Gen. USAF, Deputy Director of Plans, DCS/P&O, to AF/RDQ, Subject: *For Comment MENS—AASF and ATSF,* May 20, 1981.

[22]Hardison, DUSD Tac. War. Prg., to SAF Research, Development and Logistics, Subject: *Mission Element Need Statement (MENS) for a New Fighter Aircraft,* Jan. 11, 1982.

[23]Ferguson, P. C., *Advanced Tactical Fighter/F-22 Annotated Chronology* (Draft), Office of History, Aeronautical Systems Center, Wright–Patterson AFB, OH, Aug. 1996.

[24]Tremaine, S. A., ASD Deputy for Development Planning, to Industry Participants, Subject: *Request for Information (RFI) for Advanced Tactical Fighter (ATF) Mission Analysis,* May 21, 1981.

[25]Reis, R. J., Contracting Officer, HQ AFSC, to Industry Addressees, Subject: *Request for Information for Advanced Tactical Fighter Mission Analysis (RFI-ATF/MA-001),* May 21, 1981 (with attached RFI definition).

[26]Lowndes, J. C., "Defense Studies Specialized Aircraft," *Aviation Week and Space Technology,* Oct. 5, 1981, pp. 81, 84, 85.

[27]Memorandum of Agreement Between Air Force Wright Aeronautic Lab. and Aeronautical Systems Div. Deputy for Development Planning, Subject: *Annex 2: Advanced Tactical Fighter (ATF) Program,* Sept. 1981.

Chapter 3

PHASE 0: CONCEPT DEFINITION

OVERVIEW

As noted previously, the status of the program at the end of 1981 was that Milestone Zero had just been approved, but fiscal year 1982 funding had been denied. However, an unfunded request for information (RFI) had been released during 1981, and this provided a process for industry participation in the ATF program at no direct cost to the government. For approximately the first year after Milestone Zero approval, Advanced Tactical Fighter (ATF) activity therefore focused on managing the RFI.

Air-to-air and air-to-surface missions were given equal emphasis in the RFI. The seven participating contractors submitted a total of 19 designs, including dedicated air-to-air, dedicated air-to-surface, and multimission fighters. Industry mid-term briefings were presented to the Air Force in November 1981, and final results in May 1982. The Air Force completed its analysis of the results and issued its final report on the RFI in December 1982.

By that time the determination was made that the ATF would be an air superiority fighter with only secondary air-to-surface capability. The next generation of Soviet fighters (MiG-29 and Su-27) offered near-equality with the F-15 and F-16, respectively. New Soviet fighters also incorporated look-down/shoot-down fire control capability, which would prevent low altitude penetration by North Atlantic Treaty Organization (NATO) strike aircraft. The new Ilyushin Il-76 Mainstay (Fig. 7), commonly known as the Soviet Union Airborne Warning and Control (SUAWACS) aircraft, would further enhance the effectiveness of their improved fighter force. As a result, it became apparent that the greatest need faced by the United States was for an advanced air superiority fighter.

This need for an enhanced air superiority fighter was reflected in the program management directive (PMD) for the concept definition (CD) phase, which was circulated for comment in early 1982 and finally released on August 24, 1982. This PMD changed the name of the program element from combat aircraft technology to ATF Technologies and es-

Fig. 7 The Ilyushin Il-76 Mainstay Soviet Union Airborne Warning and Control (SUAWACS) aircraft would further enhance the effectiveness of their new generation fighter force.

tablished air superiority as the primary mission of the ATF. Two projects were established: the Advanced Tactical Fighter, which included concept and technology development for the ATF, and the Joint Fighter Engine, an engine technology demonstration program to be managed jointly in accordance with a memorandum of understanding (MOU) that was being worked out with the U.S. Navy.

These activities got under way during 1983. A concept development team (CDT) was established within the Air Force Systems Command's Aeronautical Systems Division at Wright–Patterson Air Force Base, Ohio, to manage this effort. Pratt and Whitney and General Electric each received contracts of $202 million in September 1983 for 50-month ground demonstrator engine efforts. Concept development investigation (CDI) contracts of approximately $1 million each also were awarded in September 1983 to seven airframe companies. One important change was made during the solicitation process for this effort: an increased emphasis was placed on low observables (LOs). The CDT was made aware of progress to date in the various LO programs, and special access security provisions were set up so that LO technology could be properly assessed in the ATF CDI effort.

The CDI contracts were completed in 1984. In late 1984 the ATF program office prepared for a Milestone I review, which would authorize release of a request for proposal (RFP) for the Demonstration/Validation (Dem/Val) phase. It was planned that three or four contractors would continue in this phase, which would concentrate on ground demonstrations of critical technologies, rather than on the development or prototyping of specific aircraft designs.

However, the RFP was put on hold for approximately one year because of concerns about the projected cost of the ATF. When it was finally released in late 1985, a $35 million unit flyaway cost goal had been imposed. The RFP was also modified shortly after release to strengthen the emphasis on LOs. In early 1986, while the source selection was in progress, a Presidential Blue Ribbon Commission on acquisition management released a report that recommended the use of proto-types prior to entering full-scale development of major weapons systems. A decision was made in May to include a "best-effort" prototype flight test effort in the Dem/Val phase of the ATF program. Revised proposals were requested from the contractors and were submitted in July.

The prototyping decision limited the number of Dem/Val prime contracts to two because of the high cost of developing and flight-testing the prototype aircraft. Teaming was therefore officially encouraged in order to keep more of the airframe companies involved through the Dem/Val phase. The Air Force source selection was completed in October 1986: Lockheed and Northrop were each awarded contracts of $691 million. Lockheed teamed with Boeing and General Dynamics, whereas Northrop teamed with McDonnell Douglas, to accomplish the next phase of the ATF program.

REQUEST FOR INFORMATION RESULTS

Seven companies had responded to the ATF RFI: Boeing, General Dynamics, Grumman, Lockheed, McDonnell Douglas, Northrop, and Rockwell. RFI mid-term results were presented by industry during November 1981, almost concurrently with the Milestone Zero approval. Final briefings were presented from May 17 to 20, 1982. Throughout the RFI process additional meetings and briefings were held as needed to exchange necessary information. In recognition of the need to provide a long lead time for any engine development that might be necessary, the major engine companies (Allison, Garrett, General Electric, Pratt and Whitney, and Teledyne/CAE) were invited to attend all government technical briefings to the RFI airframe contractors.[1]

Responses to the RFI varied widely, both in the characteristics of the aircraft concepts that were presented and in the quality and types of information furnished in response to the questions posed in the RFI. Air Force analysis of the RFI responses took longer than anticipated; Air Force feedback to industry briefings were held on November 16 and 17, 1982, and the government ATF RFI final report was completed by Aeronautical Systems Division Deputy for Development Planning (ASD/XR) on December 21, 1982.[1]

REQUEST FOR INFORMATION CONCEPTUAL ADVANCED TACTICAL FIGHTER DESIGNS

The seven participating companies submitted a total of 19 conceptual designs, which are illustrated in Figs. 8 and 9.[1] They ranged from a Northrop lightweight "cooperative fighter" that was smaller than an F-16 to a Lockheed "battle cruiser" bearing a distinct resemblance to the SR-71.[2] In addition, there was an "in-house" design for a subsonic low observable fighter provided by the Air Force Flight Dynamics Laboratory (AFFDL). This was a surrogate for a contractor design (submitted by General Dynamics) that could not be used in the RFI because of special access security restrictions on low observables technology. Most of the contractor-proposed designs were multirole-capable, but for the purposes of sizing, 12 were sized by air-to-air mission requirements (Fig. 8) whereas the remaining seven were sized by air-to-surface mission requirements (Fig. 9). Specific mission profiles and ground rules were selected by the contractors, so that it was difficult to draw any general conclusions from the contractor design submissions alone.[1]

Aeronautical Systems Division (ASD) therefore selected, for further analysis, four concepts that were representative of the range of designs submitted:

1) *Numbers Fighter (N):* This concept was a very lightweight, austere, low-cost design. The lower capability was intended to be offset by the larger numbers that could be procured.
2) *Supersonic Cruise and Maneuver (SCM):* The SCM design emphasized maneuverability and specific excess power in the transonic and supersonic regimes.
3) *Subsonic Low Observable (SLO):* This "in-house" concept relied on low radar cross section (RCS) and infrared (IR) signatures, rather than on classic fighter performance.
4) *Hi-Mach/Hi-Altitude (HI):* This design was intended to exploit the regime above Mach 2 and 50,000 ft.

The four concepts are illustrated in Fig. 10. Sizing studies were performed on these generic concepts for both the air-to-air and air-to-ground missions. It was found that the air-to-surface mission could best be performed by the SLO fighter, with a takeoff gross weight (TOGW) of approximately 75,000 lb. [However, if its performance did not make it unpopular with Tactical Air Command (TAC), its acronym would!] The best air-to-air concept was an SCM aircraft with reduced signatures and a TOGW of 55,000 lb.[1] The combination of reduced signatures and supercruise (to exploit the "high-fast" sanctuary) would

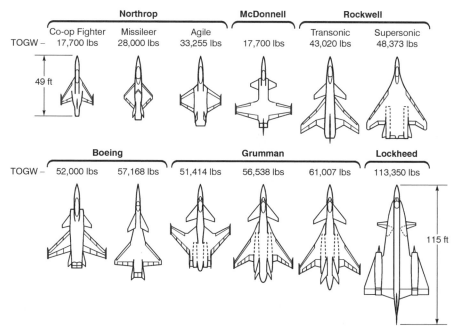

Fig. 8   RFI design concepts sized for air-to-air missions.[1]

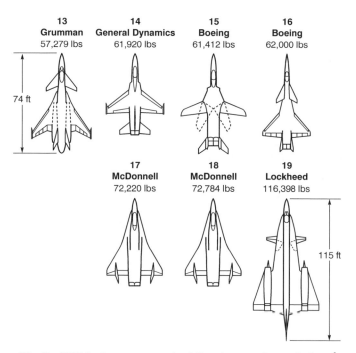

Fig. 9   RFI design concepts sized for air-to-surface missions.[1]

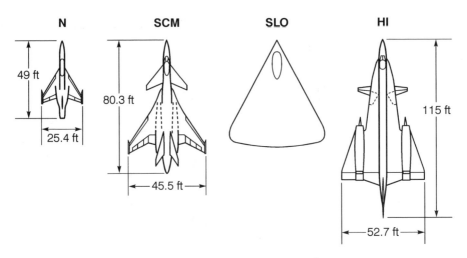

Fig. 10   Generic ATF concepts.[1]

greatly reduce the lethal zone of most surface-to-air missiles (SAMs), as it had for the SR-71, and put the ATF completely out of reach of the shorter-range systems. Mach 1.4 to 1.5 and a cruising altitude from 50,000 ft to as high as 70,000 ft were seen as appropriate values.[2,3] There was still considerable uncertainty (in 1982) as to what levels of signature reduction would be achievable in a high maneuverability supersonic fighter aircraft.

GENERAL ADVANCED TACTICAL FIGHTER CHARACTERISTICS

RFI responses tended to emphasize air-to-air considerations in the design of an ATF. Beyond-visual-range capability was strongly desired for an air superiority ATF, although the RFI responses did not specifically address issues of how that would be achieved [for example, BVR identification friend or foe (IFF) considerations].[1] However, there were extensive efforts under way in the beyond-visual-range identification area, especially focusing on noncooperative techniques, under the auspices of the Combat Identification Systems Program Office (CISPO). Information on these efforts was made available to the aircraft companies.[4]

All of the designs included short takeoff and landing (STOL) capability, with thrust vectoring and reversing.[1] The need for STOL came from the threat of runway denial in Central European scenarios. It was believed that if runway crews could fix 2–3 craters per hour and if the ATF could take off and land in 2000 ft, then enemy attacks on runways might only degrade the ATF's sortie generation rate (SGR) by around 20% (Ref. 3).

All but one of the designs used two engines, and all but one had two-dimensional exhaust nozzles. All had a single crew member. The designs offered significant improvements over the F-15 and F-16 in mission radius and supersonic dash/cruise/maneuver capability. The performance improvements were enabled by the following technologies and design considerations:

1) higher engine thrust-to-weight ratios, achieved through the use of higher engine temperatures and other improvements such as fewer stages for reduced engine weight;
2) design optimization for supersonic persistence (including super-cruise for several of the designs);
3) designs sized to more ambitious mission requirements (i.e., many designs were somewhat larger than typical existing fighter aircraft);
4) weapon–aircraft integration; and
5) composite materials.

Responses noted the need for avionics improvements in the following areas, although there was little quantitative analysis of the benefits: 1) advanced radars, 2) air-to-air fire control, 3) avionics integration—hardware integration (shared antennas/apertures, databases, computers) and functional integration (increased automation, presentation of higher order processed information, rather than raw data, to the flight crew)—and 4) stealthy avionics systems. Evaluators noted that more rigorous analysis would be needed in order to make, and justify, intelligent avionics design and technology decisions.[1,5]

There was an identifiable range of weapons loads that characterized most of the designs. For air-to-air missions, the designs carried between 6 and 12 air-to-air missiles of mixed type (short, medium, and long range). In addition, there was consensus that a gun should be included to "preclude a close-in sanctuary" for enemy aircraft (such as that enjoyed at times by North Vietnamese aircraft during the Vietnam War). Air-to-surface designs tended to have a design weapons load between 6000 and 8000 lb, although TAC later criticized this as "barely adequate" for a swing capability on an air-to-air ATF and inadequate for an aircraft whose primary mission is deep strike/interdiction.[6] There was consensus that a high degree of weapon–airframe integration would be required to achieve desired penetration speed/range performance, as well as reduced signatures. Air Force evaluators concluded that weapons integration would require additional technology emphasis if an ATF were to be developed for the air-to-ground missions.[1]

Contractors were also asked for life cycle cost estimates and inputs to Air Force cost models. Responses in this area were generally incomplete and/or too inconsistent to be of use.

RESPONSES TO SPECIFIC ISSUES

Contractor responses to the nine specific RFI issues varied widely. In some cases there was a consensus; in other cases the responses were too few, insufficiently substantiated, and/or not consistent enough to draw any conclusions. Results are summarized in the following[1,5]:

1) Aircraft/weapons performance trades: No conclusions possible.
2) Quality vs quantity: Four contractors responded to this issue. Consensus was that an ATF design should avoid the extremes of a simple, limited capability system produced in very large numbers, or a high-end, Mach 2.5-plus aircraft available in only limited numbers. However, all of the designs submitted by Northrop were near the first extreme, whereas the designs submitted by Lockheed both represented the second extreme; so it may be inferred that the consensus did not represent the views of those two companies at that time.
3) Delaying vs destroying enemy second echelon forces: No conclusions possible.
4) One vs two crew members: Generally one if primary mission is air-to-air and two if air-to-ground.
5) Base level maintenance: Specific supportability improvements were identified: onboard auxiliary power unit (APU), built-in-test (BIT) systems, and modular components with unambiguous fault detection and isolation capability. Nevertheless, consensus was that it would not be possible to eliminate intermediate-level maintenance within stated technology available dates.
6) Observables/countermeasures tradeoffs: No conclusions possible.
7) Specialized vs multirole design: A single basic design was preferable to specialized designs for each role. However, in some cases this truly meant a single aircraft for all missions, whereas other contractors proposed an aircraft that would be missionized on the production line with specialized avionics and/or weapons systems, resulting in two dedicated configurations.
8) Air base attack vs air-to-air kills: No responses to this issue were provided.
9) Impact of concepts on propulsion system requirements: A new engine would be required. However, there was little consensus on the specific characteristics of such an engine.

One of the major responsibilities of the Air Force Wright Aeronautical Laboratories (AFWAL) in the RFI process was to arrive at a technology plan to ensure that that the required technologies would be available in time for a full-scale development (FSD) decision, which was then expected at the end of fiscal year 1987. Here are the key findings of this part of the effort[1]:

1) A large amount of technology development would be necessary to support the proposed ATF design concepts.
2) The specific supporting technology plan for ATF would be dependent on a final decision on the mission of the ATF (air-to-air or air-to-ground).
3) Most existing and programmed air vehicle technology efforts were geared toward a supersonic cruise and maneuver fighter. This was consistent with what would be needed to develop an ATF for the air superiority role, although additional effort would probably be required to deliver the technology in time for the planned ATF FSD entry.
4) Most existing avionics technology programs were geared to low-level strike in night/weather conditions. Significant redirection would be necessary in avionics technology if air superiority became the primary role of the ATF.

SUMMARY OF THE REQUEST FOR INFORMATION

As noted previously, mid-term briefings were presented by industry from November 16 to 19, 1981, and final briefings from May 17 to 20, 1982. In addition to those organizations directly involved in the management of the RFI, these presentations were attended by representatives from the Air Staff, the Air Force Systems Command Headquarters (HQ AFSC), and the Tactical Air Command Director of Requirements (TAC/DR). Results were subsequently briefed to essentially all Air Force organizations that would become players in the ATF program. The final report on the RFI, including the Air Force evaluators' analysis of the industry responses, was completed in December 1982. By this time, a decision had been reached on the primary mission of the ATF.

SHIFTING EMPHASIS: AIR SUPERIORITY BECOMES PRIMARY ADVANCED TACTICAL FIGHTER MISSION

Various considerations pushed the ATF toward a pure air-to-air fighter during 1981–1982, as the RFI dialogue was in progress. Several emerging systems would at least partially fill the need for an advanced strike/interdiction aircraft. At the same time, the fielded U.S. air superiority fighter, the F-15, had grown 10 years older since the ATF was first proposed, and several new air-to-air threats were on the horizon.

AIR-TO-GROUND DEVELOPMENTS

During the late 1970s to early 1980s, two of the industry participants in the ATF program—McDonnell Douglas Aircraft and General Dynamics—began to promote improved versions of their existing fighters, the F-15 and F-16, respectively, for the deep strike role. If one or both of these aircraft could satisfy the requirements for a future strike aircraft, then the best course of action for the Air Force, and the best business plan for the companies, would be to focus the ATF on the air superiority role.

McDonnell Douglas began development of a demonstrator for an air-to-ground variant of the F-15 during the late 1970s using company funds. This aircraft, seen in Fig. 11, first flew in July 1980. It was a two-

Fig. 11   A McDonnell Douglas-developed F-15 ground attack demonstrator aircraft, based on the two-seat F-15B, first flew in July 1980.

seat, all-weather aircraft with terrain-following capability for low-level penetration. This was demonstrated to the Air Force, which expressed interest in an air-to-ground version of the F-15, although not exactly the one that was demonstrated.[7]

General Dynamics, in conjunction with NASA, first began to study a "cranked-arrow" variant of the F-16 in 1976. The original purpose was to achieve supersonic cruise capability for the air-to-air interceptor mission. It was referred to as the Supersonic Cruise and Maneuver Prototype (SCAMP) and/or the F-16XL. In early 1981, two F-16 airframes, one a single-seat F-16A and the other a two-seat F-16B, were turned over to General Dynamics for modification.[7]

By this time another role for the F-16XL had been identified; the new configuration allowed a large number of weapons to be carried semiconformally along the very large wing root. The reduced drag of this type of weapons carriage, together with the increased internal fuel load (resulting from a longer fuselage, and a much larger wing area, than a basic F-16), made the F-16XL a promising design for a strike aircraft. In this context, the two-seat F-16XL was sometimes (erroneously) referred to as the "F-16E." This was the designation reserved for a possible follow-on production variant of the aircraft that would have been developed from the F-16XL effort. The F-16XL (Fig. 12) first flew in July 1982.[7]

In 1981, even before the first flight of the F-16XL, the Air Force began a formal evaluation and comparison between two-seat F-15 and F-16 derivatives for the strike mission. Following a short flight evaluation, the F-15E was eventually chosen. FSD of the production variant was begun in May 1984, and the first production deliveries took place in 1988. The F-15E featured a high-precision, synthetic aperture air-to-ground bombing radar, an improved head up display (HUD) and bombing computer, a broad range of air-to-ground conventional and precision weapons carriage and delivery capabilities, conformal fuel tanks for increased range, and improved performance engines and strengthened landing gear to accommodate the F-15E's greatly increased takeoff gross weight. The F-15E also provided significantly enhanced night attack capability against small tactical targets using the low-altitude navigation and targeting infrared night (LANTIRN) system (Fig. 13).[7]

Thus, during 1981–1982, the time that the ATF mission was being decided, the selection of an "-E Fighter," an F-15 or F-16 derivative for the strike/interdiction role, was in progress.

Another new strike aircraft was nearing operational capability in 1982, although at the time this was only known to a few Air Force plan-

Fig. 12   The General Dynamics F-16XL flew for the first time in July 1982. A development of the two-seat version, to be designated F-16E, was proposed for the Air Force ground attack mission.

Fig. 13   Production F-15E fighter equipped with LANTIRN pods.

ners. Nevertheless, those few were in positions to influence the course of the ATF program. The F-117A (Fig. 14) had been in full-scale development since 1978. It first flew in mid-1981, and its signatures were far lower than even the "stealthiest" concepts proposed in the air-to-surface (ATS) or "1995 Fighter Study" efforts. By 1982 it had been decided that roughly three squadrons of F-117As would be procured, and the new aircraft was seen as a powerful force multiplier. Not only would it be able to attack heavily defended targets, but it was ideally suited to the mission of destroying enemy air defense sites, which would in turn allow other friendly strike assets to operate much more effectively.[8]

Finally, as a result of various improvements, it was believed that the F-111 would continue to be effective through the late 1990s, contrary to concerns expressed in the 1971 ATF CONOPS that the F-111 was already entering obsolescence at that time.[3,9]

Fig. 14 The F-117A "stealth fighter" was developed as a dedicated night ground attack aircraft. It achieved initial (but still limited) operational capability in 1983.

AIR-TO-AIR THREAT DEVELOPMENTS

However, the airborne threat was still present and becoming progressively more serious:

1) The Su-27 and MiG-29, discovered to be under development in 1978, could potentially match the capabilities of the F-15 and F-16. The follow-on generation of Soviet fighters would be superior to the U.S. fighters.
2) Soviet production rates had traditionally been much higher than Western rates for comparable systems. If the Soviets gained technological near-parity coupled with their traditionally large numbers, NATO would no longer be able to achieve and maintain air superiority.
3) Look-down/shoot-down capability in Soviet fighters (also revealed in 1978) threatened to deny low-altitude penetration to NATO strike aircraft.
4) Newer Soviet fighters showed a trend toward increased range/ loiter time (relative to earlier Soviet fighters), providing capability to loiter at altitude where they could detect and destroy any low-flying U.S./NATO aircraft attempting to penetrate Soviet-controlled airspace.[10]
5) Soviet ground forces possessed increasing deployed SAM capability to complement the Soviet fighters' look-down/shoot-down capability by denying medium and high penetration options.[11]

The air superiority challenge would be compounded by the introduction of a Soviet AWACS-type aircraft, the Ilyushin Mainstay Il-76 derivative. This was determined to be under development in 1979–1980 and would help the emerging generation of Soviet fighters to be used with maximum effectiveness.[12] Neutralizing the Soviet AWACS capability became a high priority, but this would be an extremely difficult task. The mission would require penetration of enemy-controlled airspace to attack a target that the enemy would make every effort to protect. Because the target was the enemy's most capable airborne radar system, it would be impossible for a conventional fighter to achieve any kind of surprise. The mission seemed most appropriate for a low observable fighter. A quick-look assessment conducted in the summer of 1982 determined that the F-117A, even if it were equipped with air-to-air missiles, would not be capable of performing this mission effectively.[8] The ATF would be the first aircraft with the potential to neutralize the Soviet AWACS capability through the combination of low observables and an effective long-range offensive air-to-air capability.

All of these developments suggested that the greater long-term need was for an air superiority fighter rather than for a completely new advanced strike fighter. It was believed that ongoing developmental programs, such as those that led to the F-15E and the F-117A along with upgrades to other existing systems (such as adding LANTIRN capability to the F-16), could address the perceived shortfall in strike capability, *particularly* if the ATF could provide air superiority and, specifically, protection from the coming generation of Soviet fighters. The ATF, in turn, would have the best chance of achieving that goal if it were developed as an uncompromised air-to-air fighter; the Air Force did not want to repeat some of the unfavorable experiences of the multirole Tactical Fighter Experimental (which evolved into the F-111).

PROGRAM DIRECTION AND EARLY REQUIREMENTS EVOLUTION

These considerations had an impact on the ATF program as formal guidance was developed for the concept exploration phase. Under then-current acquisition procedures, the next step following approval of a mission element need statement (MENS) was the development of a program management directive (PMD). Accordingly, the Directorate of Operational Requirements under the deputy chief of staff for Research and Development, U.S. Air Force Headquarters (USAF/RDQ) developed a draft PMD in early 1982 and circulated it to appropriate Air Force organizations for comment.

This draft PMD followed the pattern of the RFI and of the recently approved MENS in that it covered air-to-air and air-to-ground mission areas without specifying whether it was intended to develop separate aircraft for the two roles or to fill both with a single aircraft. However, TAC comments on the draft PMD strongly urged that it was time to make a determination on this issue:

> We have reviewed the new ATF PMD and are concerned with implications of a single ATF aircraft. We do not believe that the requirements for our next generation aircraft can be properly satisfied by one aircraft and are therefore drafting statements of need [SONs] for individual A-A and A-G ATFs.... [TAC also began drafting a Preliminary System Operational Concept (PSOC) for the ATF. The SON and PSOC were required acquisition documents.]
>
> The specific purpose of the PMD should be the development of next generation tactical fighters (plural). If this PMD is to address only one of these aircraft, then that fact should be clearly stated and the relationship between the programs defined....
>
> The IOC for each ATF [A-A and A-G] should be given.[13]

Subsequent dialogue clarified the intent that first priority should be given to an ATF for the air superiority role [1993 initial operational capability (IOC)], with an air-to-ground ATF to follow (1995 IOC). Air-to-ground deficiencies were still considered to exist, particularly in the area of night/adverse weather deep strike missions, which had to attack Soviet second-echelon forces before they engaged as well as hit command and control centers and supply points. NATO depended on being able to attack such targets from the air in order to counter the Soviet numerical advantage on the ground. The F-15E or F-16E was only seen as an interim solution. However, the air-to-air deficiency had higher priority because "air superiority is a prerequisite to any effective air-to-ground tactical air operations."[6]

Following the modification of the PMD to place highest priority on an ATF for air superiority, RDQ was still apparently open to the possibility of a single ATF with air superiority as its primary mission and a secondary strike capability. Again, TAC voiced strong objections, stating that the only way to satisfactorily address both deficiencies was to develop two distinct aircraft:

1) A true air superiority fighter with secondary air-to-ground capability would not adequately fill the air-to-ground deficiency.
2) Attempting to incorporate a true deep-strike capability (with an adequate payload) would excessively compromise the primary mission capability of an air-to-air ATF.

TAC agreed that an air-to-air aircraft and an air-to-ground aircraft would share many common characteristics because of the common geographical, operational, and threat environments in which both would operate. TAC also agreed that both would need some multirole capability to adapt to changing mission needs (in the long term as well as during the course of any single conflict), but insisted that this should not, in either case, be allowed to compromise the primary capability of the two ATFs that would have to be developed.[6,14]

TAC also expressed concerns about the way requirements and ATF system characteristics were being approached: some specified performance figures were too low and/or misdirected; other requirements were being specified too early, without a sound basis. For example, an RDQ White Paper on the ATF noted that air-to-air armament would consist of 8–12 air-to-air missiles and a 25- or 30-mm gun (characteristics apparently derived from the contractor RFI design submissions). The TAC position was that it was premature to specify either the number of missiles or the gun caliber:

1) Missile payload should remain open to trade studies of payload vs aircraft size. A larger number of missiles might maximize the lethality of an individual ATF, but a smaller payload per aircraft might result in a more affordable ATF, hence the ability to field a larger number of aircraft. The smaller aircraft might also be more survivable.

2) Gun caliber should be optimized in the context of the aircraft-gun combination. For example, the weight of a 30-mm gun would result in a loss of maneuverability, and possibly a *lower lethality* in air-to-air combat, than a 20-mm gun on a lighter, more agile fighter, which could track its target more effectively. In defensive maneuvering the larger and heavier gun would clearly be a liability, with no advantages to offset its size and weight penalty.

The TAC concerns were summarized as follows[14]:

> ATF development must be predicated on the threat, yet we find scant threat based rationale in many ATF discussions. ASD/ AFWAL briefings often center on new technologies.... We believe that ATF aircraft designs and schedules must be threat based with every performance parameter justified as superior to threat capabilities.
>
> The roles and missions of ATF aircraft should be those of the aircraft they replace—the F-15 and F-111. Several contractors, in responding to the ATF RFI [which was nearing completion at the time], have been unwilling to design aircraft that directly engage the projected threat for the 1990s. A predominant belief is that the low (and possibly medium) altitude environment will be "closed off" to tac air. This rationale has spawned a series of high altitude, high speed designs [the "HI" fighters, one of the generic classes which came out of the RFI responses] which seek to overfly the threat. Missing from these designs are rational concepts for weapons, seekers, and tactics. Other concepts are high stealth, low performance aircraft [the "SLO" fighters] which delay detection but offer little capability once detected. ATF aircraft cannot avoid the conflict but must go where the targets are and beat the enemy where he is found.

Relative to stealth, it should be noted that there were key people at TAC at this time who were intimately aware of the status and accomplishments of the low observables programs and knew just how well "stealth" was being made to work in the F-117A. At the same time, they also knew that the ATF could not just be another F-117A; it would have to be able to engage and defeat any projected threat in the highly competitive air superiority arena.

There was consensus during this period on several points, including the following:

1) Central Europe was the most *critical* scenario, but actual conflict was more *likely to occur* elsewhere; therefore, the ATF would need range, flexibility, and a reduced logistic footprint for more effective/responsive deployment and employment worldwide.
2) Appropriate avionics and software standards would have to be implemented, and strong attention to supportability in these areas would be necessary. A computer resources working group would need to be formed to develop the necessary support and configuration control planning documents. Technologies for improved avionics performance and supportability, such as very high-speed integrated circuit (VHSIC) modules, would need to be specifically addressed.

Revisions were completed, and the new PMD was issued on August 24, 1982. The new version, PMD R-Q 7036(5)/63230F, included both air superiority and strike/interdiction missions but identified air superiority as the first priority. This PMD also changed the name of the funding line from Combat Aircraft Technology to Advanced Tactical Fighter (ATF) Technologies, with two projects:

1) *Advanced Tactical Fighter:* Concept and technology development for the ATF, including contracted concept studies by industry. The formation of a concept development team was directed to manage this activity.
2) *Joint Fighter Engine:* An engine technology demonstration program, starting with individual component technologies and culminating in the ground testing of complete demonstrator engine(s). This project was to be managed jointly in accordance with a memorandum of understanding being developed with the U.S. Navy.

In addition to these two projects, the PMD noted that "The ATF program will also provide timing and capability focus for technology base development activity in avionics, weapons and weapons integration."[15]

ADVANCED TACTICAL FIGHTER CONCEPT DEVELOPMENT INVESTIGATION

PREPARATION

In accordance with the PMD, the ATF concept development team (CDT) was established under ASD Deputy for Development Planning (ASD/XR) in October 1982. Maj. Claude Bolton, who had been assigned to ASD/XR to head the ATF effort in the summer of 1982, became the first CDT director. The primary mission of the CDT was to prepare for (and subsequently manage) the funded contractor concept development studies. The CDT started out small, approximately half a dozen people, and contained, from the beginning, an unusually large percentage of personnel with sustainment and logistics backgrounds.

In late 1982, Congress appropriated $23 million (of a requested $27.3 million) for the ATF program in fiscal year 1983. This was the first significant funding for ATF. Almost immediately, however, it was noted that proposed funding cuts contained in the fiscal year 1985 Air Force Program Objective Memorandum (POM) could delay the ATF program, slipping IOC from 1993 to 1995.

An announcement of the pending ATF concept definition studies appeared in the *Commerce Business Daily* on November 2, 1982. Based on the just-completed Air Force analysis of the RFI responses, a draft RFP was prepared and was in circulation for comment by January 1983. A draft Tactical Air Forces Statement of Operational Need (TAF SON) was issued on February 13, 1983.[16]

As the ATF program began to enter the systems acquisition process, information on the program was disseminated through many channels. General officers representing various organizations in the Air Force received summaries of the mid-term and final RFI results during two Next-Generation Fighter Conclaves hosted by the ASD Commander on February 9, 1982, and August 19, 1982, respectively. (Lt. Gen. Lawrence Skantze was ASD commander at the time of the first Conclave, and Lt. Gen. Thomas McMullen at the time of the second one.) RFI "Feedback to Industry" meetings were held on November 16 and 17, 1982. OSD also received briefings on the ATF program during late 1982.

In October 1982, Frank Campanile [formerly Aeronautical Systems Division Directorate of Mission Analysis (ASD/XRM) chief analyst on Advanced Tactical Attack System Mission Analysis (ATASMA) and subsequently involved in Advanced Tactical Fighter Mission Analysis (ATFMA), the XRM mission analysis effort that proceeded in parallel with the RFI] presented "Potential Attributes of the Next Generation Fighter" at the Society of Automotive Engineers Aerospace Congress. In March 1983, the ATF program figured prominently in an AIAA Aircraft Prototype and Technology Demonstrator Symposium (held at the Air Force Museum adjacent to Wright–Patterson Air Force Base). An entire session at this symposium was devoted to technology thrusts and the ATF. The session featured a panel discussion moderated by retired Lt. Gen. James T. Stewart (a former ASD commander) on the use of prototypes and technology demonstrators

in the ATF program and closing remarks by Lt. Gen. Thomas H. Mc-
Mullen (ASD commander).

Members of the Air Force Scientific Advisory Board (SAB) received
briefings on various aspects of the ATF program during late 1982 and
early 1983, culminating in a SAB summer study on the ATF program in
August 1983. Also during the summer of 1983, Lt. Gen. McMullen as-
sembled a General Officer Steering Group for ATF Technologies. The
purpose of all of these activities was to build widespread support for
the ATF program and to ensure that all relevant concerns could be
brought to light and addressed as early as possible in the acquisition
process.[1,16]

INCREASED EMPHASIS ON LOW OBSERVABLES

During the latter half of 1982 and early 1983, it was determined that
the progress that had been achieved in low observables needed to be
made available to the ATF to ensure that the ATF would be able to
successfully meet the threats of the 1990s and beyond. Brig. Gen.
Bolton recently recalled, "That was about the time that someone
grabbed us as we walked down the hall and said 'We need to talk to
you.' "[17] Bolton and other members of the CDT were made aware of
the existing low observables programs [managed under ASD/XRJ, the
Directorate of Low Observables, which in spite of its symbol did not report to the deputy for Development Planning (ASD/XR) but instead
reported directly to the ASD commander] and the special access required (SAR) security requirements that went along with low observables technology. At around that time ASD/XRH, the Directorate of
Design Analysis, formed a "Low Observables Unit" in order to "get ed-
ucated in the area of low observables and to pursue the subject of reducing observables on air vehicles."[18]

Signature experts from the "black world" began to look at how the
technology could be applied to a high-performance fighter aircraft and
how much signature reduction might realistically be expected. However, by the time any kind of a plan to properly address low observables (LO) technology in the ATF program could be formulated, the
CDI RFP was already out.

The RFP for the ATF concept development investigation (CDI) had
been circulated in draft form during late 1982 through early 1983, and
the "final" RFP was released to industry on May 18, 1983. The Joint Advanced Fighter Engine (JAFE) RFP was also released in May 1983 to
General Electric, Pratt and Whitney, and Allison. The stated due date
for the CDI proposals was June 17, 1983. The requirement for increased
emphasis on LO appeared as an amendment to the RFP on May 26,

just eight days after the original RFP was released. The amendment required additional low observables tradeoff data to be developed and was accompanied by a three-week extension on the proposal due date.[16]

Lockheed and Northrop, the two companies that were most heavily involved in the ongoing low observables programs, had reportedly been lobbying for an increased emphasis on low observables in the ATF.[19] Up until late 1982 or early 1983, certainly through the completion of the RFI process, the ATF program had been proceeding independently of the various covert programs that were developing true "very low observable" aircraft—Have Blue, Senior Trend (F-117), Tacit Blue, Advanced Technology Bomber (B-2), etc. Signature reduction was considered for ATF, but not generally to the extent that was being achieved in those programs. The increased emphasis was intended to ensure that the full range of available LO technology was properly assessed in an integrated fashion.

PROPOSALS AND CONCEPT DEVELOPMENT INVESTIGATION CONTRACT AWARD

Proposals in response to the amended RFP were received on July 8, 1983. Technical evaluation of the proposals was performed between July 11 and July 29. Contract negotiations were then conducted, and the CDI contracts were actually awarded on September 2, 1983.[20] Seven companies—Boeing, General Dynamics, Grumman, Lockheed, McDonnell Douglas, Northrop, and Rockwell—received contracts of just under $1 million each. Concurrently, special access security arrangements were made with each contractor to ensure that adequate facilities and procedures were in place to handle the low observables portions of the studies. The kickoff meeting was held on September 14. The studies were to last eight months with an interim report due in December 1983 and final results to be presented in May 1984.[16,21]

TRANSITION TO A SYSTEM PROGRAM OFFICE

During the spring and summer of 1983, the ATF concept development team (CDT) began to expand in preparation for the increased level of activity as the funded studies got under way. Col. Albert C. Piccirillo was assigned as the new CDT director and ATF program manager on June 16, 1983. Previously, he was responsible for analyzing advanced aircraft, weapons, and employment concepts as chief of the Tactical Systems Division in the Air Force Center for Studies and Analyses (AF/SA). Once Col. Piccirillo was assigned to the ATF Program, the CDT began to be referred to as a "SPO Cadre," the core of what would become a new System Program Office (SPO). The CDT was formally redesignated as an SPO in October 1983, although this did not in any way change the direction of the program. ATF remained under ASD/XR at this time.[16] The primary tasks for the SPO would be to[3,22]

1) manage ATF concept development activities by the potential prime contractors;
2) monitor the work of the Joint Advanced Fighter Engine (JAFE) engine contractors (in conjunction with ASD/YZ, the new engine SPO, which managed the JAFE effort) and insure airframe/engine compatibility;
3) coordinate and integrate all other ATF-related critical technology and subsystem development;
4) develop an Air Force ATF concept and preliminary specification; and
5) prepare for the Milestone One review by the Defense Systems Acquisition Review Council (DSARC).

The intent from the start was to develop a weapon system that would meet the intent of the users' requirements and offer as many desired capabilities as possible while avoiding "requirements of marginal value and disproportionately high cost."[3] Stanley A. Tremaine, the ASD deputy for Development Planning, stated "that designers are very serious about design costs, with affordability an important parameter in ATF design."[23] In response to questions about what an ATF could do that existing fighters could not, Maj. Gen. Robert D. Russ (director of Operational Requirements under the deputy chief of staff for Research, Development, and Acquisition) replied that "part of the reason for getting started is to be able to tell you the answer to just that question."[3]

Requirements were not yet firm at this stage. Consistent with the RFI results, the CDI studies were oriented toward the "Supersonic Cruise and Maneuver" class of fighters, with additional emphasis on low observables. The CDI RFP called for performance that was consistent with this concept. However, the RFP did not specify "hard" requirements, but rather ranges of performance parameters, and asked for tradeoff data within the ranges specified. Deviations above or below the specified ranges were even permitted, if justified by significant payoff in other areas and/or enhanced total system effectiveness. "At the end of the Concept Development phase, however, the Air Force plans to have a very much more firmed up requirement."[21]

Much of the original basis and advocacy for the supersonic cruise requirement for ATF had its origins in the results of a national "Conference on the Utility of Supersonic Cruise," sponsored by ASD/XRH (the Directorate of Design Analysis) and held at Wright–Patterson Air Force Base in 1976. John Chuprun, the director of Design Analysis, was the guiding force behind the conference; he devised it, chaired it, summarized the conclusions, and provided the rationale for a dry power supersonic cruise capability in the next-generation U.S. fighter. Later, in conjunction with the release of the draft PMD in early 1982, XRH formed a dedicated ATF team led by Mr. Wayne O'Connor to provide broad design tradeoffs for TAC's Directorate of Requirements (TAC/DR) to illuminate the impact of technology and performance variations on overall ATF system cost. The initial set of design trades was accomplished in time to influence preparation of the draft TAF SON, which was issued in February 1983.

The initial set of performance goals established by TAC were based solely on projections of Soviet threat capability with only limited analytical insight into their effect on aircraft size or cost. In-house ASD design trades accomplished by XRH showed, for example, that one of the more than 20 TAC-specified maneuver goals (sustained g at supersonic, high altitude conditions) was causing an inordinate impact on both ATF gross weight (and cost) and on the required engine cycle. TAC/DR agreed that the capability added by this particular maneuver level was not sufficient to justify its extreme cost impact and reduced it prior to the CDI phase. Reduction in this single requirement to a more realistic level saved over 10,000 lb in gross weight and 10% in cost and allowed the ATF engine to be designed for much more efficient operation.

When it was determined in 1983 that low observables were important to the ATF, XRH was tasked with reviewing all available information on low observables programs and technology and incorporating low observables considerations into the design tradeoffs. This resulted in briefings to TAC on the impact of reducing signature to different levels on the size, cost, and performance of the ATF. These studies also showed the limits to signature reduction allowed by technology and which technologies drove the conclusions, allowing decisions to be made on where risk reduction efforts should be focused. These studies were critical because they showed what could be done with full knowledge of all highly classified ongoing low observables activities, which no individual aerospace company had at that time.

Another significant in-house tradeoff study done during this time period by XRH was an ATF payload trade. Internal carriage of weapons in a supersonic aircraft can result in significant size and cost penalties because of the sensitivity of the design to added volume. There was a desire from an operational point of view to provide the maximum internal radar-guided missile capability possible, but the penalty was great due to the large span of the basic AMRAAM fin design. It was shown that a specific reduction in missile fin span would allow six AMRAAMs to be carried in the same internal weapons bay space that would otherwise be occupied by only four missiles. This gave rise to development of a reduced-span AMRAAM variant early enough to allow its use in subsequent F-22 aircraft detailed design development.

ADVANCED TACTICAL FIGHTER AS ENVISIONED DURING
CONCEPT DEVELOPMENT INVESTIGATION

The ATF at this time was understood to be an air superiority fighter incorporating supersonic cruise and maneuver capability, reduced observables, long range, STOL capability, and improved reliability/maintainability/supportability (relative to current-generation fighters), as described in the following contemporary (October 1983) article published in *Flight International:*

> "There is a lot of emphasis on long range in ATF," says Col. Piccirillo. There is also a major emphasis on supportability, so that the aircraft can operate from austere, dispersed airstrips in Central Europe and elsewhere, including the Middle East. ATF must generate high sortie rates to meet a numerically superior threat....
>
> After analyzing [the RFI responses], the Air Force concept development team at Aeronautical Systems Division has drawn up a preliminary statement of its requirements, along with detailed 1990s threat scenarios and representative ATF mission profiles. This baseline specification has been issued to industry....
>
> The baseline specification is broad.... Where a performance parameter is specified, so is a range of values over which designers can perform trade-off studies to determine, for example, the best speed and altitude capability. The aim is to produce an integrated design with balanced capability, providing not only performance but survivability, availability, and supportability.
>
> Survivability is vital, says Col. Piccirillo, if ATF is to re-establish air superiority over hostile territory. The aircraft will combine integrated electronic warfare with signature reduction to enhance survivability.... Weapon carriage will be integrated with the airframe at the earliest design stage to reduce drag and radar cross-section....
>
> The Air Force has not yet decided whether ATF will be a single-seat or two-seat aircraft....[24]

However, it was generally believed that the ATF would be single-seat. This would require an unprecedented degree of cockpit integration.[25]

COCKPIT AND AVIONICS INTEGRATION. Up until the Vietnam War, fighter aircraft had progressively increased in complexity, and responsibility for managing the various systems had been left to the crew. The ATF would be even more sophisticated, but the lessons of Vietnam were that the flight crew's workload would have to be reduced. Therefore, the ATF's cockpit would have to be designed to present information in a concise, integrated way and to make the flight crew's tasks more manageable. Brig. Gen. Ronald W. Yates, the F-16 program manager at the time, stated:

> Cockpit integration and simplification are what we need. We have got to the point where we provide the pilot with tremendous amounts of information, highly compressed, from many sources, all in real time.... He doesn't need any more sensors or switches. He needs something to tell him, "Here are your targets and your threats, and this guy is going to kill you unless you deal with him now."[26]

Sensor fusion (the presentation of information from multiple sensors on a single display), various avionics integration concepts, voice control, and other technologies were expected to be used to reduce the workload and improve the effectiveness of the crew.

It had reportedly been decided by that time that the ATF would carry an internal gun, and for defense would carry expendable countermeasures dispensers and onboard electronic countermeasures (ECM) equipment. Integrated Electronic Warfare System (INEWS) was specifically under consideration; Col. Piccirillo met with Col. Kelly [Aeronautical Systems Division Electronic Warfare System Program Office (ASD/RWW)] in September 1983 to establish the interface between the ATF and INEWS programs. The INEWS concept development RFP was due to be released in November 1983. It was reported that "tradeoffs are currently being made on such parameters as antenna location vs aerodynamics and maintenance requirements of the ATF."[21] A tri-service beyond-visual-range combat identification system and the Joint Tactical Information Distribution System (JTIDS) were also under consideration at that time.[21]

SUPPORTABILITY. The ATF should be "designed for the crew chief when it is on the ground," according to a TAC officer quoted in early 1984:

[TAC] wants the new advanced tactical fighter designed from the start for ease of maintenance.... The command is working closely with Air Force Systems Command and contractors. Part of this effort is to demonstrate turning around an aircraft such as the F-16 for additional sorties so that contractors can observe the operation under realistic conditions and use the data to design the new fighter. A systems approach will be taken to designing the advanced tactical fighter to reduce required airlift to support operations with the aircraft and reduce the required flight line support units....[10]

The following ATF supportability goals were consistently reported during the period of the CDI contracts[10]:

1) Complete *a full combat turnaround* in less than 15 min.
2) *Break rate* should be less than or equal to 6% (break rate is defined as occurring when an aircraft "completes a mission and lands with a nonfunctioning system that will prevent completion" of its subsequent assigned mission).
3) *Fix rate* should be greater than or equal to 75% with a 4-h period with an 85% fix rate being achieved within 8 h.
4) No more than 2% of the aircraft should be down awaiting parts at any one time.

These and other goals were developed very early in the concept definition phase of the ATF program. Gen. William Creech, the commander of TAC, expressed some concern during 1983 that the goals were too ambitious and that it was premature to specify hard numbers. The ATF CDT worked closely with the TAC Directorate of Operational Requirements (TAC/DR) to resolve this and other issues, as the statement of operational need (SON) for the ATF was being worked through several drafts at that time.[27] Nevertheless, the values stated above are essentially representative of what appeared in the SON when it was finally released in late 1984.[28]

TECHNOLOGY INTEGRATION. Regarding the wide range of technologies that would be needed to achieve the capabilities envisioned for the ATF, the report was made that "the tough part will lie in sorting out the technologies, and in forming the right combination at the right time for incorporation in the ATF."[26] Lt. Gen. Thomas H. McMullen, commander of ASD, noted that "the integration of technologies has become a technology of its own."[26] From the aerospace companies' point of view, "85% of what each of us [the companies] is doing is pretty well known to all the others. But the other 15% is highly proprietary, and a lot of it has to do with how we plan to put all the technologies together."[26]

In recognition of the need to mature various technologies in parallel with the progression of the ATF through the acquisition process, a technology demonstration component of the ATF program was formally established by a new version of the program management directive that was issued in November 1983. PMDR-Q7036(6)/63230F added the critical technology demonstration project to the ATF program, along with the two existing projects (ATF concept development and Joint Advanced Fighter Engine). The critical technology demonstration project (described in Chapter 5) did not lead directly to any specific technology demonstrations, but it did help formulate an overall technology plan for the ATF.

Effort was made in the ATF program to benefit from lessons learned on prior fighter development programs, and in particular the F-15. Only about a week after Col. Piccirillo's arrival at ASD, there was a two-day meeting with Maj. Gen. Ben Bellis, who had been the program manager in the early phases of the F-15, and other F-15 program veterans to discuss issues, problems, and lessons learned from that program.[4] One of the important lessons was in regard to reliability, maintainability, and supportability (RM&S). The F-15 had been designed with more emphasis on those qualities than its predecessors (and incorporated, for example, a high degree of accessibility to subsystems). However, "the driving element behind the F-15 development effort was performance, with supportability being a secondary consideration. This is being changed with ATF," according to a late 1983 press report. "Supportability, maintenance, system safety and simulator design are being built into the system at the early stages in order to reduce costs down the road. Industry has been asked to report not only on initial costs, but on LCC [life cycle cost] as well."[21]

The F-15 had been stuck with one very difficult challenge in the area of RM&S: the engine. The Pratt and Whitney F100 engine for the F-15 was designed primarily to achieve a then-unprecedented thrust-to-weight ratio of around 8 to 1. The F-15 and its F100 engine had been developed in a relatively short time period in the early 1970s, with virtually no Dem/Val phase, in order to counter the advanced Soviet threat then being developed, the MiG-25. Durability and reliability were, to some extent, sacrificed in order to achieve the engine's performance goals within the abbreviated time frame. The reliability problem was then aggravated by the F-15's unprecedented climb, acceleration, and maneuvering performance, which could cycle the engine through the full range of operating conditions much more rapidly than earlier fighters. Pilots, of course, liked to put the F-15 through its paces, resulting in very high failure rates for the engines. To avoid repeating this situation, durability and reliability of the ATF engine would be emphasized, and the duration of planned ATF engine ground testing would be "five to six times longer than in past programs," according to a November 1983 interview with Col. Piccirillo. The total amount of testing prior to FSD actually ended up being 30 times as many hours than on the F100![29]

The ATF engines would be designed from the ground up, not only to meet the performance goals of the ATF, but also to achieve improved RM&S by incorporating lessons learned from operational experience with the F100/F110 generation of afterburning turbofan engines. Additionally, various programs expected to develop and demonstrate avionics concepts and technologies for the ATF were also under way at this time. (These are discussed in greater detail in Chapters 5 and 6.)

In conjunction with the initiation of engine development, it was recognized that a flight demonstration of thrust vectoring would be important. Accordingly, a STOL demonstrator program was formulated, expected at the time (late 1983) to be either an F-15, F-16, or F/A-18 with two-dimensional vectoring nozzles. "The two-dimensional nozzles are currently on ground test, and contracts should be let soon for flight test, scheduled for 1987," said Col. Piccirillo. "Experience will feed directly into ATF, reducing the development risk."[24] An RFP for the thrust vectoring demonstration was released in September 1983, approximately concurrent with the contract awards for the advanced engine demonstration programs. In 1984, McDonnell Douglas was picked to modify an F-15 for the demonstration. This became the F-15 STOL/Maneuver Technology Demonstrator, developed from 1984 to 1988 and flight tested from 1988 to 1991, which provided valuable experience and inputs for ATF.

PROGRAM PLANS AS OF 1983–1984

The plan to produce 750 ATFs at a peak rate of 72 per year emerged at around this time, although questions were raised very early as to the realism of such a plan. The F-15 production rate had never exceeded 42 per year. If ATF cost estimates were based on a production rate of 72 per year, and this rate were never achieved, then unit costs would increase.[3] In early 1984, this was reported:

> Given the technological unknowns—and a Congress eager these
> days to suggest less expensive alternatives—many observers, both
> in the Defense Dept. and in industry, wonder whether the ATF's

proponents are being far too ambitious. "The Air Force already is taking the high-risk road on the plane," warns one Pentagon insider who predicts that the price tag will eventually dominate all deliberations over the superfighter. Even if all these new technologies mature in time, he adds, trade-offs are likely because no one can afford to put as many radical developments into one aircraft as the service wants.[25]

On more than one occasion, cost concerns led to studies aimed at reducing the size, weight, and cost of the ATF. As design concepts were narrowed down following the RFI studies, analyses had indicated that the ATF would have a maximum takeoff weight between 55,000 and 60,000 lb. In March 1984, this was reported, "The Air Force is asking industry to consider tradeoffs in technology in reducing the advanced fighter from a 55,000-lb design to a 45,000-lb one."[10]

The concept development studies by the contractors were completed in May 1984. Following an analysis of results by the Air Force, the plan was to select three or four contractors for a competitive demonstration and validation (Dem/Val) phase. Consistent with the strategy endorsed at Milestone Zero, the Dem/Val phase would focus on the development and demonstration of technologies rather than on the construction of high-fidelity prototypes. Some anticipated program events (and actual funding amounts) were as follows[24]:

1) First funding was requested in fiscal year 1983 [$23 million[30]].
2) Thirty-five million dollars was approved in fiscal year 1984 [$37.4 million was requested; the House Armed Services Committee (HASC) recommended zero, but the Senate Armed Services Committee (SASC) supported ATF[23]].
3) First substantial funding was received in fiscal year 1985 [$99.6 million—$15.2 million concept development, $75 million ATF Engine, and $9.4 million critical technology demonstration[29]].
4) Early 1985, the field was to be narrowed to two or three companies for Dem/Val.[25]
5) FSD "of a single design" was scheduled to start in 1987.
6) IOC was anticipated for 1993.[29]

As of late 1983, the Five-Year Defense Plan (FYDP) included $99.6 million in fiscal year 1985, $250 million in fiscal year 1986, $275.4 million in fiscal year 1987, $351.7 million in fiscal year 1988, and $1.1 billion in fiscal year 1989. However, the Air Force fiscal year 1985 budget request (submitted in late 1983) reflected a modified schedule, with FSD start in fiscal year 1989 and IOC in 1995. Under the modified plan, fiscal year 1989 funding would be $650 million rather than $1.1 billion. This would better match the available funding level and also allow more time for the integration of the many new and emerging technologies expected to be used by the ATF.[1,29]

On May 4, 1984, the announcement was made that the ATF would eventually be maintained and supported by the Air Logistics Center depot at McClellan Air Force Base in Sacramento, California. Gen. James P. Mullins, commander of the Air Force Logistics Command, stated that the assignment was made this early in order "to ensure that all support functions are considered during the design and development phase" of the ATF.[31,32]

DESIGN EVOLUTION DURING THE CONCEPT DEVELOPMENT INVESTIGATION

In general, many of the CDI design concepts closely resembled those that had been presented in the RFI responses.[17] The Lockheed approach is of some interest, because this company eventually became the prime contractor for the F-22. During the CDI phase, Lockheed shifted from their initial "battle cruiser" concepts (large, fast, high-flying aircraft as typified by their RFI designs) to a series of very low observable designs that were directly descended from the F-117A. However, these were considered by the Air Force to be very limited in traditional aspects of fighter performance, and the Air Force was adamant that the ATF must be an outstanding air superiority fighter— as reflected in the TAC comments quoted earlier.

The design that Lockheed had submitted in response to the CDI RFP looked like a larger and longer F-117 with several major differences. These had included a high mounted wing, four tails (instead of the F-117's two tail configuration), and inlets that were placed below and behind the leading edge of the wing. The highly faceted design had serious aerodynamic problems and weighed about 80,000 lb. Bart Osborne, Lockheed's ATF chief engineer during this period, recently noted:

> It had become clear that ATF was going to be superstealth and not a cousin of the YF-12 or the SR-71. I stopped the YF-12 derivative effort, and we started working on an F-117 derivative for ATF.... We knew we would have serious problems with the supersonic requirements. Our design could go supersonic, but it was a real dog of an airplane. With enough power, you can make a brick fly. We did not know how to analyze a curved stealthy shape in those days. The software wasn't sophisticated enough, and we didn't have the computational capability we needed. We felt our hands tied by the

analytical problems. Lockheed had become convinced that if we could not analyze a design as a stealthy shape then it would not be stealthy. We would not break that barrier until 1984.[33]

By mid-1985 Lockheed had begun to evolve a design that combined a reasonable size, low observables, and the types of performance that the Air Force was seeking. Lockheed's Bart Osborne again noted, "We simply started drawing curved shapes, even though we could not run the designs through our analytical software models. When we went to curved airplanes, we began to get more acceptable supersonic and maneuver performance. Instead of relying on software models, we built curved shapes and tested them on the company's radar range. The curved shapes performed quite well in the radar tests." The last major challenge to replacing facets with curved surfaces had involved the radome. Recalled Osborne, "We didn't know until early 1985 how to make a stealthy curved radome. We started drawing them in late 1984 before we knew how to analyze them."[33] By early 1985, the ATF Lockheed design incorporated fully curved surfaces. This design continued to evolve and eventually would emerge as Lockheed's Dem/Val proposal design, which is described in a later section. Teaming arrangements made later in the program would further influence the design, which would ultimately become the F-22, as it is recognized today.[34]

PROGRAM MANAGEMENT

The ATF program was structured in accordance with then-current acquisition reform activities. Steve Rait, the chief contracting officer in the ATF SPO, reported in 1984:

> To achieve affordability, we in the Advanced Tactical Fighter System Program Office, Wright–Patterson Air Force Base, Ohio, have enthusiastically adopted the Department of Defense Streamlining Initiative....
> Former Deputy Secretary of Defense Paul Thayer signed, on January 11, 1984, a memorandum to secretaries of the military departments calling for improvement in Department of Defense contract requirements. The memorandum contains recommendations that "call for precluding untimely, untailored and accidentally referenced application of specifications and standards and for specifying 'results' required rather than detailed 'how to' procedures in contracts and requests for proposals."[35]

The Streamlining Initiative was based on the following tenets:

1) Utilize contractor ingenuity and experience.

2) Encourage early industry involvement; for example, utilize draft RFPs.
3) Specify what is needed, not how to do it.
4) Specify system level functional requirements early.
5) Require contractors to tailor for the coming phase of a program.
6) Avoid premature applications of military specifications and standards.
7) Limit contractual applicability to one level of references.
8) Pursue economically producible, operationally suitable, and field supportable designs.
9) Assure complete production specifications while providing the contractor with flexibility to optimize the design.

In keeping with the recommendation to involve contractors early, internal business planning meetings were held in the spring and summer of 1984, along with meetings with each potential prime contractor in April 1984, to "develop mutually acceptable and understandable business approaches."[35] On October 16, the SPO released 150 copies of the draft Dem/Val RFP to potential prime contractors, interested subcontractors, and government agencies. Comments were requested, particularly with regard to "specifications, standards and requirements which may be overly restrictive or costly." Comments were due on November 13 to support final RFP release, expected at that time to be in December 1984. Over 1400 comments were received.[35]

The original draft had been designed to comply with the spirit of the Streamlining Initiative. For example, emphasis was placed on what was needed, not how to accomplish it. Because the ATF would represent such a large technological jump relative to earlier systems, plagiarizing from past program documents did not prove practical, therefore affording an opportunity to start from the ground up and avoiding the accidental application of inherited specifications and standards. Of those referenced, many were cited as guidance only rather than as a contractual requirement. Offerors were instructed to include in their proposals a description of their plans to tailor the advanced tactical fighter's specifications.[35]

However, the comments received illustrated one of the difficulties of accomplishing acquisition reform—it is very difficult for the many institutions involved to agree on what constitutes an improvement in the acquisition process:

> Not everyone is advocating the streamlining initiative. We received approximately 30 draft request for proposal comments recommending application of *additional* Department of Defense stan-

dards and specifications. Suggestions to tailor were, however, many and varied.... Several agencies noted that our D/V [Dem/Val] requirements were too detailed for a demonstration/validation program, and that specific goals stated in the draft request for proposal were too ambitious and unnecessary. One comment that indicated the Streamlining Initiative is not yet institutionalized came from a government contract administration activity. We state in the request for proposal that specific (many) specifications are cited for guidance only. The suggestion was to make them contractual in the D/V phase to "hold the contractor's feet to the fire."[35]

TRANSITION TO DEMONSTRATION AND VALIDATION

PREPARATION FOR MILESTONE I AND DEMONSTRATION AND VALIDATION REQUEST FOR PROPOSAL

As noted previously, final briefings and reports had been delivered in May 1984 by the seven contractors involved in the concept definition contracts. Air Force analysis of the results continued through the summer of 1984. In July 1984, ATF program oversight responsibility was transferred (within ASD) from the deputy for Development Planning (Stanley Tremaine) to the deputy for Tactical Systems (Brig. Gen. Gerald C. Schwankl),[36] This move was consistent with its progression from the conceptual stage toward development and demonstrations.[29]

An announcement of the coming ATF demonstration and validation contracting activity was made in the *Commerce Business Daily* on October 2, 1984. On October 16, a meeting was held with the ATF CDI contractors for the dual purpose of debriefing them on the Air Force analysis of the CDI reports and providing them with the draft RFP for the Dem/Val phase.[1,35] In addition to the seven prime contractors, it was anticipated that draft RFPs also would be provided to approximately 60 government offices and 50 interested subcontractors (which it was, as described). The final RFP release would be made after the Milestone I decision, which was expected in December 1984.[36]

MILESTONE I DELAY

The Defense Systems Acquisition Review Council (DSARC) Milestone I review was scheduled for December 7, 1984. Pre-briefings were conducted with various Air Force organizations throughout October and November. Working-level meetings and pre-briefs were also held with various Office of the Secretary Defense (OSD) offices. These briefings were generally well received and did not precipitate any major changes.[37]

On October 31, the Air Force Cost Analysis Improvement Group
(AF CAIG) stated that cost estimates for the ATF appeared reasonable at this stage of the program. In November the System Threat
Analysis Report (STAR) was validated by the Defense Intelligence
Agency. On November 6, the Air Force Council received the Milestone
I pre-briefing, and no significant changes were directed. On November
9, the Air Force Requirements Review Group validated the statement
of need (SON) for the ATF. At a pre-brief the same day to the Assessment Committee of the Air Force Systems Acquisition Review Council (AFSARC), however, it was stated that "There was concern that
showing the projected high unit ATF costs to Congress at this time
could jeopardize the current (good) level of support for the program."[38]

The AFSARC Milestone I review—the highest level Air Force review, prior to the final DSARC review—was held on November 13, and
as a result of this meeting, the Council adopted two goals intended to
keep the ATF affordable: a takeoff gross weight goal of 50,000 lb and a
unit flyaway cost goal of $40 million (FY85 dollars).[16] At that time Col.
Piccirillo noted, "The goals lie at the lower bound of the ranges which
we have shown in our SCP [System Concept Paper].... We are re-
working our in-house 'notional' design and cost around these objectives and will include them in revisions to our DSARC briefing per Dr.
Cooper's direction.... [Dr. Thomas Cooper was the Assistant Secretary of the Air Force for Research, Development and Logistics
(SAF/AL)]."[39]

The DSARC review, however, was subsequently canceled, only
about a week before it was scheduled to occur.[40] Verne Orr, secretary
of the Air Force, had indicated two areas of concern regarding the ATF
program:

1) *Affordability*—"We must be able to show that we can meet our
   overall force structure requirements while remaining within the
   funds that are likely to be provided to the tactical position of our
   budget," assuming that the total defense budget as a fraction of
   the gross national product, and the Air Force's tactical modernization funds as a fraction of the defense budget, would not increase within the foreseeable future.[40,41]
2) *Close Air Support*—"What is the Air Force doing about Close Air
   Support?" An Air Force commitment to provide "enhanced Close
   Air Support" to the Army "was one of the '31 initiatives' agreed
   to by the Army and Air Force Chiefs in a May 84 Joint MOA
   [Memorandum of Agreement]. Mr. Orr is apparently concerned

that the Air Force is not adequately addressing this area," and would not have the funds to do so if it made a commitment to start Dem/Val of the ATF at this time.[41]

Secretary Orr wanted to be sure that these concerns were addressed before proceeding with the DSARC review.[41]

Briefings were prepared for Secretary Orr on these issues, and concurrently the SPO revised its RFP, source selection plan (SSP), and other program documentation to comply with the guidance that followed the November 13 AFSARC brief. By February 20, these revisions were complete. From February through the spring of 1985, the ATF SPO awaited Secretarial approval to go to the DSARC and to release the Dem/Val RFP.[42]

Meanwhile, on February 1, 1985, the TAC preliminary system operational concept (PSOC) for the ATF was published. The PSOC stated:

> The description of the ATF is based on a set of characteristics from each of the seven contractors' preferred designs [from the recently completed concept development contracts]. Because of the similarities of these designs as they affect operations and maintenance, separate preliminary SOCs will not be required. The ATF will be employed in numbers and missions similar to those used in fighter operations today; however, it will be much more capable of performing offensive counter air missions and surviving. Reliability and maintainability will be major driving influences on ATF design, and will result in a fighter aircraft that can maintain the high [sortie generation rate] SGR required in wartime operations.[43]

The PSOC also noted that the ATF would be a single-seat fighter, that a major emphasis would be on neutralizing the enemy's look-down/shoot-down fighter threat, and that operational requirements had been traded extensively against cost to achieve an affordable weapon system.

The need for the ATF was not in serious doubt. In April 1985, the leaders of TAC and the major commands of the Tactical Air Forces reviewed the ATF program "against the framework of the total fighter roadmap" and presented the following conclusions to Gen. Gabriel, the Air Force Chief of Staff:

> A) The ATF requirement is based primarily on the other side's advances in high performance fighters and look-down/shoot-down capability. The key operational characteristics which derive from the requirement are fundamentally sound.

B) Our growing investment in air-to-ground capability, the need
to make that investment survivable, and our important role in support of air-land battle scenarios emphasize even more the importance of providing effective air superiority at times and places of
our choosing, hence ATF.

C) Affordability is a must for ATF. While the performance and
R&M [reliability and maintainability] requirements are about right
for this stage of the program, we stand ready to continue the requirements trade-off process as cost drivers are identified in order
to field an affordable fighter....[44]

Regarding the continuing delay in the preparations for Dem/Val, the
group noted:

We need to proceed with release of the RFP for the Dem/Val
phase soonest in order to protect the program schedule...industry
is finding it difficult to keep a quality commitment on the ATF in
the face of the current uncertainty surrounding the program and
other programs competing for the same talent....

While ATF ranks very high in our priorities already, to reinforce
our commitment to it, we will be raising it even higher as we rack
up our TAF [Tactical Air Forces] requirements later this year. We
urge you to prevail on the secretary once again and we stand by to
echo the TAF's strongest unified commitment to ATF.[44]

At the same time as the recommendations previously noted, Gen.
Skantze (AFSC commander) made the following statements to Secretary of the Air Force Verne Orr:

We are prepared to solicit proposals for the demonstration and
validation phase of the Advanced Tactical Fighter program. I realize that you have great concern regarding the cost of the ATF, and
so do I. Our objective, as stated at the November 1984 AFSARC
[Air Force Systems Acquisition Review Council], is to build a system which will satisfy the Tactical Air Force's basic requirement and
do it for a unit recurring (flyaway) cost of $40M (FY85). We will
continue to work with the operating commands to scrub the requirements to determine what level of capability can be delivered.

If we proceed with some dispatch into the demonstration and
validation phase, complete our DSARC I, and award contracts by
December, we should finish the requirements scrub [review] by
September 1986....[45]

However, there was growing controversy over ATF costs. In an April
1985 report on Air Force tactical force budget issues, the Congressional

Budget Office (CBO) stated that several cost assumptions related to the ATF were unrealistic. Apparently, flyaway cost figures as low as $30 million had been quoted in "informal conversations" between Air Force officials and the CBO. At the time F-16 and F-15 flyaway unit costs were approximately $15 million and $25 million, respectively.[46]

Congress was also concerned about whether the necessary technology would be available in time for full-scale development of the ATF, particularly in the area of avionics integration. OSD voiced concerns about whether there was sufficient attention to supportability in the ATF program. Derivatives of the F-15 and F-16 continued to be proposed as cheaper alternatives to the ATF.

Development costs as well as unit flyaway costs were a source of concern. In May, the SPO was tasked by HQ USAF to identify possible ways of reducing the total projected ATF Research, Development, Test, and Evaluation (RDT&E) costs to three possible levels: $8 billion, $9 billion, and $10 billion. The SPO estimate at that time appears to have been slightly over $10 billion (TY dollars), while the contractors were briefing much lower estimates in Washington. Eliminating areas of conservatism in the cost model, reducing capability at IOC, and accepting a higher degree of program risk were all to be considered as possible ways of reducing the projected cost. Slipping the IOC was also considered at this time.[47]

Concurrently, Secretary Orr tasked AF/SA (the Air Force Center for Studies and Analysis) to conduct a "fiscally constrained" study of the ATF. These were the guidelines for this study:

1) "Air Force will continue to receive (for the next 20 years) the 1986 dollar allocation plus inflation."
2) Probable ATF cost should be based on "straight line extrapolation" from historical fighter cost data (both RDT&E and unit).
3) "Using a historical basis, estimate the percentage of the Air Force budget which may be devoted to TAC Air."
4) Show what the production of 300 ATFs would involve relative to these constraints.
5) Show impact to other tactical programs, including strike, CAS, and tactical airlift.

Finally, Orr noted, "I am not particularly interested in studies which are premised on an artificially based cap, such as a fighter not-to-exceed 40,000 lb, or a fighter not-to-exceed $40 million...."[48] Col. Piccirillo noted that "straight line extrapolation" correlated very well with historical trends in fighter costs and that a first cut using the ASD cost database predicted a unit flyaway cost of $43 million (FY86 dollars) for 300 ATFs.[49]

Preliminary results of the AF/SA study were shown at an ATF working group meeting at the Pentagon on June 19. Col. Piccirillo noted,

> The basic projections, which are based upon data which does not include some advanced, twin engine sophisticated aircraft such as the F-18 and F-14, indicate a historically projected ATF unit flyaway cost of $25–36 million as being possible. Backup analyses, including the F-18 and F-14 data, give projected costs roughly equivalent to the SPO estimates of $40–42 million for unit flyaway cost. Our [the ATF SPO's] concern was expressed over presenting these possibly misleading lower estimates to Mr. Orr, which could lead to a flyaway cost goal even lower than the current $40M being imposed...."[50]

In July 1985, the AF/SA final results were presented to Orr and Cooper, who stated, "It is now appropriate to proceed with our request to enter the Dem/Val phase." A review to the AFSARC principals was scheduled for mid-to-late August, to be followed as soon as possible thereafter by the DSARC.[51]

At that time the ATF program received some questionable attention from the OSD Supportability staff, who submitted a proposal to the Defense Resources Board, which would "require the Air Force to create a new 'ATF R&M project line' to do ATF Dem/Val supportability demonstrations on a long list of OSD-prepared projects which they believe the Air Force laboratory/tech base is not adequately supporting."[52] There were 16 projects on the list, including active array radar antenna, engine digital electronic controls, modular avionics, environmental control system/avionics cooling technology, testability, connectors, and maintenance in a nuclear/biological/chemical (NBC) threat environment. Many of these areas were already being worked in the ATF program and/or related technology programs. Nevertheless, the proposal, which was supposedly supported by Wade, the assistant secretary of Defense for Acquisition and Logistics, would have added $208 million under a separate project to work those specific efforts:

> These would then be individually tracked by the OSD staff to determine progress in improving R&M prior to a milestone II ATF decision.
>
> The Air Staff has been resisting this OSD initiative, especially since sources for funding of this magnitude have not been identified [i.e., some or all of it could end up coming out of ongoing ATF

activities], and also due to the potential precedent for even more DOD micro-management down to subsystem level. However, all of the above seems to indicate that the ATF program is getting more support than is healthy from all of the diverse special interest groups in Washington.[52]

The proposal was not adopted. Initiatives such as this often arise simply because some outside agency is not aware of what is already going on in a program. Wade did eventually receive a briefing on ATF supportability and was impressed with the approach that the program was taking.

Also in July 1985, the Presidential Blue Ribbon Commission on Defense Management (the Packard Commission) was created to examine practices and recommend improvements in defense acquisition management.[16] The findings of this commission would ultimately have a significant impact on the course of the ATF program.

In preparation for the coming Dem/Val RFP release, a sources sought synopsis was published in the *Commerce Business Daily* on August 12.[53] On August 14 the new Undersecretary of Defense for Research and Engineering (USDR&E), Donald Hicks, indicated that he felt the DSARC review should wait until after RFP release when more specific information would be available (but before contract award, probably spring 1986). However, Hicks did want an information briefing. Furthermore, Cooper still desired an AFSARC principal-level review prior to RFP release.[54]

Following the usual pre-briefing cycle, this AFSARC principal review was held on August 23. The next ATF SPO weekly activity report noted:

> As a result of this review, we expect to receive direction to release the Dem/Val Request for Proposal in the form of an implementation memorandum. We understand several versions of the implementing memo have been drafted.... USAF/RDQT [the Air Staff office of primary responsibility (OPR) for ATF matters] provided us a draft copy of a proposed memo from SAF/AL [Asst. Secretary of the Air Force (Research, Development and Logistics), Dr. Cooper] to USAF/CV [USAF Vice Chief of Staff, Gen. Pietrowski] which contains specific direction which will require changes to the current RFP. The most significant of these are the change to require aircraft size compatibility with modified TAB VEE (first generation) aircraft shelters and further iterations of the R&D and production fly-away cost objectives...we are reviewing the RFP for areas which would be impacted and generating appropriate "slip page" changes.[55]

Formal direction followed in a memorandum from Cooper dated September 6, 1985:

> Affordability remains the key ATF program issue…. I am therefore requesting that we mount an all out effort, from the onset, to made the ATF program the showpiece for Air Force acquisition ingenuity. Let's call our initiative 'Project Acquisition Showpiece.' We must combat program costs on all fronts and all levels. As an initial step in this initiative, I am lowering our ATF unit flyaway cost goal from $40 million to $35 million (FY85 dollars). I am also tying this goal to a production rate of 72 per year over a total program buy of 750 aircraft. The previously established not-to-exceed takeoff gross weight of 50,000 pounds remains in effect.
>
> I request that all of the ATF program documentation (RFP, source selection criteria, etc.) be modified to reflect our commitment to cost control. I recognize that it is not prudent to set a unit cost *cap* on the program at this time and that the $35 million unit cost *goal* is somewhat arbitrary from a price versus capability viewpoint. The Demonstration/Validation phase will give us better insight into the tradeoffs that may be needed to achieve this goal. However, from an Air Force affordability viewpoint, the goal is not arbitrary…. As we do this [work the ATF down to $35 million], we must make sure that the previously stated specification requirements remain unchanged….
>
> I believe we should encourage joint ventures or teaming arrangements by the contractors in both the Demonstration/Validation and FSD phases. The Demonstration/Validation RFP should reflect this point and not be silent on the subject.
>
> Finally, it is highly desirable that the ATF should be capable of operating from the modified first generation Tab V shelter. We need to do the performance tradeoff studies to ascertain whether it is feasible to make this a hard requiremen…. I assume that the system requirements review (SRR), six months into this [Dem/Val] phase, is a good point to finalize this requirement.
>
> We have been relieved of the requirement to meet with the DSARC before releasing the ATF RFP. However, before we release the RFP, I would like to review the modifications that are required to meet the guidance set forth in this memorandum….

The ATF SPO completed the directed revisions to the RFP. Gen. Lawrence A. Skantze, commander of AFSC, wrote to the vice chief of staff of the Air Force on September 24 expressing his concern over the $35 million cost goal:

> 1) I have several major concerns regarding the actions requested by Dr. Cooper in his 6 Sept. 85 ATF action memo to you.

2) ATF formulation began in 1977, has continued through pre-concept contractor studies in 1979–82, a SAB Technology Application Study in 1983, Concept Definition in 1983–4, and to the instant milestone of being ready for Dem/Val RFP release. The program has followed a logical course to Milestone I....

3) In response to the Secretary's concern about ATF affordability, I committed on 15 May 85 to accept a $40 million cost cap—with a special system requirements review to be conducted 9 months after Dem/Val contract award. This commitment was reiterated at the 23 Aug. 85 AFSARC, but was rejected on the premise that the corporate Air Force could not control the process. F-15 and F-16 history shows that the Air Force could set, and achieve, unit cost goals—but only after defining detailed configurations.

4) There is clearly more logic to a $40 million unit cost estimate than there is for a $35 million goal. The $40M is closer to the SPO mid-price estimate. This mid-price estimate is very much just that—an estimate—since at this stage of the development process we must use generic and parametric cost estimating methodology. A set of point estimates will be possible after we have entered Dem/Val and have some specific designs to evaluate. Establishing the $35 million figure on the basis that it is necessary in order to procure ATF along with C-17, SICBM [Small Intercontinental Ballistic Missile], AMRAAM [Advanced Medium Range Air-to-Air Missile] and SRAM II [Short Range Attack Missile II], etc., is speculative at best. Out year projections by RD [Air Force Research and Development] and SA [Air Force Studies and Analyses], even at reduced budgets, indicate the acquisition can be worked. More importantly, priorities between major force programs will change with the need and the threat....

6) In Nov. 84, AFSC changed the Source Selection Criteria (AFR 70-15) to rank design reliability, maintainability, and producibility as the first factors within the top priority criterion, Tech Performance. That emphasis was embodied in the ATF RFP in Dec. 84.

7) I concur with the SAF/AL statement that our objective must continue to be a true, effective balance of cost, schedule, supportability and operational performance. In addressing the four variables in a predefinition phase, the acquisition logic which has developed over many years, and is embodied in the current DODI 5000.1, achieves the balance.... Only with the proper definition of all elements can cost trades be made. Conversely, if the unit cost is fixed first, then the logic prejudges what is affordable.... We run a high risk of degrading performance, reliability, maintainability, and life cycle cost

before knowing what the logical tradeoffs should be. Because we have a competitive environment and the contractors want to win, it is crystal clear from my experience that they will take the $35 million as a winning requirement and subopti- mize implementation of the operational requirements into an ATF configuration....[16]

On September 25, the revised RFP was shown to Cooper, together with a presentation by the Directorate for Design Analysis (ASD/XRH) on the possible capability impacts of a $35-million-unit flyaway cost goal. Cooper approved RFP release subject to one final change, to make it clear that the Air Force planned to solicit offers for FSD only from those contractors awarded a Dem/Val contract. This would not pre- clude other companies from submitting proposals or teaming between a successful Dem/Val contractor and one or more unsuccessful Dem/Val contractors; it merely stated that the Air Force would only *so- licit* FSD offers from the Dem/Val prime contractors.[56,57]

DEMONSTRATION AND VALIDATION REQUEST FOR PROPOSAL

The source selection plan and RFP release were finally approved by Secretary Orr on October 7, 1985. Copies of the RFP were provided to the seven airframe contractors on October 8. According to the press re- lease issued at the time, "The ATF demonstration/validation phase will include wind tunnel tests, subsystems tests, mock-up, man-in-the-loop simulations, and supportability demonstrations, leading to detailed de- signs suitable for full scale development."[58] Proposals were due on De- cember 10. The stated date for an FSD decision was early 1989, with first flight to be late 1991. IOC was specified as 1995. The definition of IOC at that time included 52 production aircraft delivered—4 for op- erational test and evaluation (OT&E), and 24 each for one training squadron and one operational squadron.[59]

Several organizations in the Air Force and OSD received briefings on ATF supportability during this period and were generally impressed with the approach that was being taken. Steve Conver, deputy assistant secretary of the Air Force (Programs and Budget) provided acting Sec- retary of the Air Force Edward C. Aldridge Jr. (previously the assistant secretary of the Air Force for Financial Management) with the follow- ing comments:

In general, I believe that we are putting unprecedented effort in these two areas [R&M, and Producibility]....

> R&M is being treated as a number one priority evaluation factor in the Dem/Val source selection; it is ranked equal with technical performance and cost.
>
> The TAF Statement of Need specifies numerical R&M and supportability goals. I am told this is the first time such goals have appeared in a SON. We have goals for break rate, fix rate, sortie generation rates, personnel requirements, mobility requirements, and O&S [operations & support] costs.
>
> Preliminary ATF studies have identified a number of "long pole" [the most significant] areas that represent the most likely causes of unscheduled maintenance actions. The Air Force has initiated laboratory programs to address each of these R&M "long pole" areas....
>
> The RFP requires the bidders to specify a number of items related to producibility. For example, the contractors are required to provide not only an aircraft design, but also the factory design that they would use to ensure a producible aircraft....[60]

On November 4, an ATF R&M briefing was presented to Wade, the assistant secretary of Defense for Acquisition and Logistics. Wade was likewise impressed with the R&M approach; however,

> Dr. Wade also expressed personal concern that the Air Force may have prematurely reduced the flyaway cost goal to $35 million without the full insight that Dem/Val will provide. He noted that it is essential that ATF be a very high performance weapon system to deal with future threats and supportability-related features may end up suffering if the cost goal is too low. His comment was "a $40 million ATF may be far more cost effective than a $35 million [ATF] and that should be the basis for such decisions." We reassured Dr. Wade that we require trade-off assessments during Dem/Val to develop a fuller insight prior to converging on a specific ATF design for FSD.... He also stated that he is concerned that "all of the services may be going too far on competition." He expressed concern that the services are forcing the primes "to beat up on the little guys to get unlimited data rights." He felt that this could eventually stifle innovation and ultimately damage the industrial base.... Dr. Wade concluded by asking that the Air Force come directly to him if it appears that there are roadblocks being caused by current DOD acquisition policies. He stated that he wants such impediments identified rapidly so prompt corrective action can be initiated.[61]

Congress also wondered about the rationale behind the $35 million cost goal. Shortly after RFP release, Congressional markups to the fiscal year 1986 defense budget threatened to severely reduce the available funding for the ATF program. Reasons for the cuts included "late RFP release" and "lack of definition of what the Air Force is giving up in ATF in going from the $40 million flyaway cost to $35 million." Additionally, "the program should proceed at a slower pace...no more than $100 million may be spent on the JAFE [Joint Advanced Fighter Engine] development in FY86."[62] Further markups threatened to seriously restrict the use of funds for ATF avionics integration—an area about which Congress had specifically expressed concern.[63] Ironically, on more than one occasion, Congressional concern as to whether the necessary technologies would be available in time for the ATF led to funding cuts that impacted, more than anything else, those very same critical technologies.

Col. Piccirillo expressed concern that some emerging technologies would have to be eliminated and/or that the schedule would have to be stretched, as a result of the proposed cuts. "According to Congressional staffers, however, Capitol Hill's questioning of the program was meant to signal concern that the Air Force did not have a firm grip on the ATF's schedule and costs. According to a November 1985 Senate report, 'The current program does not provide sufficient time to absorb all of the technology, and the program entails unacceptable risks.'"[3] This was one of many such "warning shots" from Congress. The Air Force in turn claimed that many of the advanced technologies on the ATF would actually *reduce* the overall life cycle cost of the system. Ultimately, the Air Force requested and obtained relief from the language that restricted expenditures on specific critical technology work.[16] However, reductions in the overall outyear funding available for ATF led to an 18-month IOC slip at this time, from early fiscal year 1995 to late fiscal year 1996 (but no reduction in capability at IOC). This restructuring was one of several options considered, some of which entailed smaller schedule slips but reduced capability at IOC.[64]

Around the same time, Congress floated the idea of combining the ATF with the Navy's Advanced Tactical Aircraft (ATA), a proposed long-range, low-observable, high payload medium-attack aircraft to replace the Grumman A-6. (The ATA later became the A-12, a model of which is seen in Fig. 15; it was canceled in 1991.) A joint Air Force/Navy panel had already studied this question in 1984 and had determined that a single common aircraft was not feasible.[16] Rather than merge the two programs, each of which had very ambitious objectives already, the Air Force and the Navy therefore agreed to consider each other's aircraft as follow-on replacements for the F-111 and the Grumman F-14, respectively (i.e., Air Force would consider ATA to replace the F-111

Fig. 15   The General Dynamics A-12 low-observable attack aircraft was intended to replace both the U.S. Navy A-6 and the U.S. Air Force F-111 in the deep strike role.

for interdiction, while Navy would look at an ATF derivative as a possible replacement for the F-14, seen in Fig. 16. It was uncertain what degree of modification would be required in either case).[3] The Navy ATF (NATF) program is discussed in Chapter 7. Furthermore, the Air Force and Navy had been studying subjects such as ATA/ATF avionics commonality since 1984.[65] During late 1985 and early 1986, Congress directed the two services to develop a percentage goal for avionics commonality between the two aircraft, looking in particular at whether a goal of 90% was feasible.[64] In early 1986, congressional staffers urged the Army's Light Helicopter Experimental (LHX) program manager

Fig. 16   The NATF was proposed as an eventual replacement for the Grumman F-14 Tomcat as the U.S. Navy standard fleet air defense fighter.

to join this effort to ensure the maximum possible avionics commonality between the LHX, the ATF, and the ATA.[66]

In late November 1985, the ATF SPO came under pressure from OUSDR&E (headed at the time by Don Hicks) to issue more stringent signature requirements, particularly in the aft sector. The original Dem/Val RFP contained what would be considered "very low observable" (VLO) signature levels at most aspects. However, somewhat higher signatures were originally permitted in the aft sector. The rationale was that it might not be realistic to expect very low signatures in the aft sector while meeting the ATF's very aggressive goals in speed, acceleration, and maneuverability (incorporating afterburning, thrust vectoring, and/or reversing). All other VLO aircraft—existing and/or under development at that time—were relatively low performance, subsonic aircraft. A high performance, VLO engine nozzle had never before been developed.

The prevailing view at the time was that signature treatment in the aft sector would entail a direct penalty in performance measures such as Ps, the specific excess power, which in turn determined the maximum acceleration that was possible at any given flight condition. There was still a lot of debate over how much signature reduction needed to be achieved in the aft sector and how much performance penalty was acceptable. Many operators wanted uncompromised performance, whereas others felt that, in order to be survivable while operating over hostile territory, the ATF should be as stealthy as possible at all aspects.

Nobody knew whether it was possible to achieve both at the same time. There were engineers at Lockheed and Northrop, the two companies with the most experience in low observables, who claimed that it could be done. In any case, it would not be accomplished if the RFP did not ask for it. Dem/Val would be the appropriate time to find out. The Dem/Val RFP was therefore modified with stronger LO goals in the aft sector, in order to be open to the possibility that very low aft-sector signatures and high performance could be achieved together and gain a better understanding of the tradeoffs that were involved, prior to entering FSD. One other important change was to require the contractors to include full-scale radar cross section (RCS) pole model testing as part of their proposed Dem/Val efforts.[4]

While the modification was being prepared, the contractors were advised, just over a week prior to the original due date of December 10, to "put their proposals on indefinite hold."[67] The RFP amendment was issued on December 16, 1985, and the contractors were asked what impact the changes would have on the time required to complete their proposals. Based on the contractor feedback following this session, the proposal due dates were extended to February 18, 1986, for the technical portion, and March 5 for the cost proposal.[4,68]

DEMONSTRATION AND VALIDATION PROPOSALS

Proposals were received by the revised due date from all seven companies who had participated in the CDI phase: Boeing, General Dynamics, Grumman, Lockheed, McDonnell Douglas, Northrop, and Rockwell.[7] As Gen. Skantze had predicted, all contractors claimed a unit flyaway cost of right around $35 million in fiscal year 1985 dollars. The respondents reportedly spent an average of $2 million each in support of their proposals; some spent considerably more.[3] (For comparison, the *contracted* effort for the concept development phase was only $1 million per company.) Evaluation of the technical proposals began promptly on February 19, with most of the SPO personnel moving to ASD's dedicated source selection facility to conduct the evaluation.[69]

SHIFT TO PROTOTYPING

ORIGINAL DEMONSTRATION AND VALIDATION STRATEGY

The Dem/Val program was initially planned to involve three to four competing prime contractors, who would not fly prototypes during this phase. This was consistent with the strategy adopted at Milestone Zero in November 1981 of "focused technology thrusts" rather than early competitive prototyping. There would be a formidable design challenge in combining very low observables, high maneuverability, and high speed, including supersonic cruise capability, in one aircraft. However, the capabilities were each individually well understood, and most of the individual technologies—including propulsion and avionics—could be worked through analysis, simulation, ground testing, and demonstration. In October 1985, Col. Piccirillo stated, "There are a lot of programs, many of them in the black world, that have given us a high degree of confidence that we can do this."[2]

To reduce risk, the Dem/Val phase would involve avionics development and test, RCS and wind-tunnel model testing, miscellaneous subsystems testing, the JAFE prototype ground tests, and aircraft and engine supportability demonstrations involving, for example, full-scale mock-ups with access panels (and the ground crew wearing full chem/bio warfare gear!).[2,3]

Other key aspects of Dem/Val phase plans and approach are summarized as follows by Steve Rait, the principal contracting officer at the ATF SPO:

> We plan to choose four of these defense contractors this year [1985] to demonstrate and validate advanced tactical fighter concepts. Development of critical subsystems will occur concurrently.... Late in 1988, we plan to select one airframe contractor or contractor team to proceed into full-scale development (FSD) of the advanced tactical fighter. First flight should occur in 1992 with initial operational capability occurring in 1995.
>
> Acquisition strategy for the advanced tactical fighter has been specifically tailored to protect the interests of the Air Force and to use taxpayer dollars most effectively.... A firm fixed-price contract type was selected in order to limit the Air Force cost risk in the environment that precedes a major down-selection; i.e., selection of an advanced tactical fighter FSD contractor. The primary incentive for contractors competing for award of the advanced tactical fighter FSD contract will be to outperform their competitors. To outperform could, and probably will, mean that contractors will "spend to win." The Air Force is unwilling to fund the difference in contract cost caused by these competitive pressures.
>
> The D/V request for proposal will include funding profiles so that contractors are aware of available funding. Contracts will include proposed warranty provisions for application in future advanced tactical fighter production contracts. Hopefully, knowledge of planned warranty requirements will positively affect the quality of aircraft and subsystem designs. Contractors will be notified that the Air Force expects to be granted unlimited rights in data and computer software within a specified period after delivery of the first production aircraft. If unlimited rights cannot be granted for certain items, contractors will be required to develop alternate mechanisms; e.g., second sourcing, to ensure that excluded items may be competitively acquired by the government. Finally, associate contractor agreements and interface control working groups will be essential requirements of advanced tactical fighter D/V contracts to achieve necessary integration between the airframe and critical subsystems.[35]

PACKARD COMMISSION REPORT AND PROTOTYPING

On February 18, the same day the Dem/Val proposals were received, the Presidential Blue Ribbon Commission on Defense Management, also referred to as the Packard Commission, delivered its interim report. Two of the most significant recommendations affecting the ATF program were:

> A high priority should be given to building and testing prototype systems and subsystems before proceeding with full-scale development. This early phase of R&D should employ extensive informal competition and use streamlined procurement processes. It should demonstrate that the new technology under test can substantially improve military capability, and should as well provide a basis for making realistic cost estimates prior to a full-scale development decision. This increased emphasis on prototyping should allow us to "fly and know how much it will cost before we buy.... "
>
> Federal law and DOD regulations should provide for substantially increased use of commercial-style competition, relying on inherent market forces instead of government intervention. To be truly effective, such competition should emphasize quality and established performance as well as price, particularly for R&D and for professional services.[70]

On April 1, 1986, President Reagan issued a directive to the services to begin implementing the commission's recommendations, which included the application of this "fly-before-buy" competitive procurement strategy in major defense programs. As the Air Force's most prominent new aircraft acquisition program at that time, the ATF naturally came under consideration for applying the commission's recommendations.[16]

Accordingly, in early April, the SPO was tasked to present three ATF options at an upcoming four-star level Air Force program objective memorandum (POM) review[71]: 1) a "fiscally unconstrained" option, maintaining the original 1995 IOC; 2) a constrained option based on the funding contained in the fiscal year 1987 President's budget; and 3) an option that included prototyping in the Dem/Val phase. Col. Piccirillo reported, "The interest in prototyping apparently is resulting from Packard Commission recommendations and appears to be gaining very strong support in the Secretariat and Air Staff." Acting Secretary of the Air Force Edward C. Aldridge Jr. was particularly supportive of this plan.

The decision was made very shortly thereafter in favor of the prototyping option, and the following press release announcing the decision was issued on May 1[4,71]:

Air Force Selects ATF Program as Acquisition Model

The Air Force will develop and acquire its Advanced Tactical Fighter (ATF) consistent with Department of Defense acquisition initiatives that have been put in place in recent years and the recent recommendations of the Packard Commission.

According to acting Air Force Secretary Edward C. Aldridge, Jr., the ATF is well suited to take advantage of the Commission's rec-

ommendations because of the advanced technologies involved, the
Air Force's commitment to tightly control costs, and the critical
need for the new aircraft....

Details of the acquisition approach will be briefed to the Defense Systems Acquisition Review Council within the next few
weeks before a milestone decision later this year.

The announcement was followed by formal direction to the ATF SPO
(an Interim AFSC Form 56) on May 6, to prepare a modification request
(MR) for the ATF Dem/Val RFP with the following provisions[72]:

1) Inform bidders that "the scope of Dem/Val will include best-
   effort flying prototypes and avionics prototypes."
2) Inform bidders that only *two* contractors will be selected for
   Dem/Val.
3) "The MR should strongly encourage teaming."
4) Request cost proposals for accomplishing the prototype effort.
5) Ask how the contractor will tailor the remainder of the Dem/Val
   efforts to optimize risk reduction within available funds.
6) Include funding profile in the MR.

The MR, and an overall revised Dem/Val plan incorporating prototyping, were developed very quickly because prototyping had been
under serious consideration before. In fact, prototyping had never been
very far from being a viable option in the ATF program. While formu-
lating the original Dem/Val strategy during the summer of 1984, the
ATF SPO had held half-day sessions with each contractor to learn their
views on what activities would be necessary during Dem/Val. Several
contractors had been strongly in favor of prototyping. There were also
communities in the Air Force, OSD, and Congress that advocated prototyping.[4,73] However, in the case of ATF, the technical justification for
it had never been quite compelling enough to justify the added up-
front funding that would be necessary.

Some of the principal arguments for and against prototyping multiple designs during Dem/Val are summarized in the following.

1) Reasons not to fly prototypes
   a) Early competitive prototyping prematurely "froze" both the
      supporting technology level and the weapon system designs.
   b) Available funding limited the Dem/Val phase to two compet-
      ing concepts, whereas three or four could be funded without
      flying prototypes. This would keep more concepts in the run-
      ning.

c) Improved analysis and ground testing techniques made it less necessary.
d) Many technical challenges were unrelated to air vehicle design (e.g., avionics).
e) VLO performance had already been proven in other programs (Have Blue and Tacit Blue demonstrators; F-117A strike aircraft).
f) Prototyping added up-front cost and time.
2) Reasons to fly prototypes
a) Prove critical air vehicle performance capabilities and validate the operational concept prior to final selection of a design.
b) If design goals are risky or ambitious, then having two teams build and fly their aircraft improves the probability that at least one will achieve the goals.

The single most central issue for ATF was whether prototyping would improve or degrade the Air Force's prospects for obtaining a fighter with sufficiently advanced capabilities to ensure total air superiority well into the 21st century. Opponents argued that "contractors will spend the next 30 months in the assembly plant…instead of in a laboratory," thereby reducing the level of advanced technology that could be included.[74] A near-term competition would cause the contractors to build the best airplane that they could build "right now" rather than looking ahead to the best airplane that could be built in 5 or 10 years. On the other hand, it has been argued that committing to FSD and production without first flying a prototype would have created the more risk-averse program and led to a lower performance aircraft. The way that prototyping was implemented in the ATF program (as will be described) was structured to avoid creating a risk-averse competitive situation. The contractors would not be competing their prototype aircraft against each other directly, but rather using their prototypes in the way each team thought best, to mitigate risks, while concurrently maturing various technologies through other Dem/Val activities.

Prototyping appeared to make even more sense as the original Dem/Val proposals were reviewed (a "quick look" evaluation briefing was presented to the ATF Source Selection Advisory Council on March 27).[75] Two of the ATF proposals stood out above the rest, so the budgetary considerations (only being able to fund the prototyping of two competing concepts) were not so much of a problem as originally thought. According to one Air Force official, "We had two excellent proposals, three good proposals, and two who just didn't quite get the idea."[3]

Rather than funding three or four competitors as originally planned, it now made sense to pick the two best, and then capture the contributions of the "good" contenders by allowing and encouraging them to team with the winners. Early prototyping would also add political security to the program; Congress would be less likely to cancel a program that had actually flown aircraft that had demonstrated and validated the bold concept upon which the ATF performance goals were based.

Under the revised plan, each airframe contractor would deliver two flying prototypes of their aircraft. These would be "best-effort" proof of concept prototypes, not intended for a direct competitive flyoff or to show compliance with every performance requirement, but rather to demonstrate that each company's concept was fundamentally viable. The ATF contractors would have maximum flexibility in determining their respective flight test plans. It was agreed early on that "the prototypes would not be used for an OT&E, but instead that the data generated from the prototype phase would be used for preliminary operational assessments and would serve to provide part of the data base for the ATF FSD decision."[76] Each airframe contractor would be required to flight-test both of the ATF prototype engine designs, the Pratt and Whitney YF119 and the General Electric YF120.[77]

The details of the revised acquisition strategy, including the RFP modification request, were briefed to the AFSARC on May 21. AFSARC approval of the plan was formally given in a memorandum from Dr. Cooper dated May 27. [In addition to implementing the AFSARC decision, this memo restated the 50,000-lb takeoff gross weight and the $35 million cost goal (FY85 dollars) based on a production run of 750 aircraft at 72 per year.] The MR was issued to the contractors on May 28. Revised technical proposals were due on July 28, with cost and contract volumes to follow on August 7.[77,78] Very shortly after the MR was released, the Air Force allocated the designations YF-22A and YF-23A to the two ATF prototypes.[80]

Negotiations with the JAFE engine contractors were also begun in early May to modify the JAFE contracts "to include development of flying prototype, JAFE-derived engines to meet the first flight of the prototype aircraft." Originally, there had been no plan to actually fly the new engines until the engineering and manufacturing development (EMD) phase. Flying during Dem/Val would require accelerated engine development and additional testing to qualify the engines for flight. Consistent with the new focus, the program name was changed from Joint Advanced Fighter Engine (JAFE) to Advanced Tactical Fighter Engine (ATFE).[72] ASD/YZ, the JAFE engine SPO, issued a "mutually agreed to" change order to the JAFE contracts on June 11. Under the revised contracts,

The competing engine contractors will provide weapon system integration efforts, increase their special technology (LO) efforts and reorganize their AMT [accelerated mission testing] test approach to provide flight clearance engines....

This change is for a limited effort and will be followed by a supplemental agreement to complete the ATF prototype engine program.[81]

The full restructuring of the ATFE contracts was accomplished through "ATFE Prototype Restructure" contract modifications with each ATFE contractor, which were signed in December 1987. A full discussion of the engine development is given in Chapter 6.

REVISED PROPOSALS AND MILESTONE I

On July 2, 1986, Boeing, General Dynamics, and Lockheed announced that they would team if any one of them were selected.[16] Each would submit an independent proposal, but if one of them were selected, that one would become prime contractor and the other two would become principal subcontractors. The winning proposal would be the basis for the team's Dem/Val effort. If two were selected, the remaining one would choose which of the successful bidders to team with. Northrop and Mc-Donnell Douglas reached a similar agreement a few weeks later.[82] Revised proposals from all seven airframe contractors, reflecting the new structure of the Dem/Val phase, were submitted on July 28, 1986.[83]

Because of both their historical interest and their relationship to the evolution of the F-22A design, the ATF Dem/Val design concepts submitted by the three individual companies that eventually formed the Lockheed ATF team are briefly described in the following section.

The Boeing ATF design (illustrated in Figs. 17–20) had a single chin inlet, a trapezoidal-shaped wing, and two canted tails, and it was optimized for higher operational speeds. The design had three phased-array radars in the nose and two wingroot-mounted infrared search and track systems to meet the 120-deg field of view requirement. The aircraft had three weapon bays: two side bays for short range infrared guided air-to-air missiles and a bottom bay for the longer range radar guided AMRAAM.[84]

Dick Hardy, the former Boeing F-22 program manager, recently noted, "Our designers argued most over two tails vs four tails. The whole Boeing company got involved in the argument. We had special teams set up to study the problem. Two tails won out. We thought we could meet all the requirements with two tails, which gave our design a lower signature and lighter weight." The single chin inlet incorporated an internal

Fig. 17   Front quarter view of the Boeing ATF Dem/Val design.

Fig. 18   Top view of the Boeing ATF Dem/Val design.

Fig. 19   Front view of the Boeing ATF Dem/Val design.

Fig. 20   Bottom view of the Boeing ATF Dem/Val design.

airflow splitter to feed the two engines. The inlet also had an internal variable ramp—necessary to achieve its higher design cruise speed.

The General Dynamics ATF design, seen in Figs. 21–24, featured a diamond-shaped wing with a serrated trailing edge, a very large single vertical tail, and side inlets. Two multifunction radar arrays were mounted above the inlets, and an infrared search and track system was located in the nose section of the fuselage. The radar beam from each array was steerable over an angle of 60 deg each direction from each array's face. This provided radar coverage from straight ahead to 120 deg aft on each side of the aircraft. The weapon bay arrangement was similar to that of the current F-22 with a lower fuselage bay for the radar-guided missiles and two side fuselage bays for the infrared-guided missiles.[85]

The wing planform was chosen to minimize weight while providing the maximum turn capability and supersonic cruise. The single vertical tail, however, presented problems in achieving a totally stealthy design. General Dynamics ran many wind-tunnel tests to find a location and shape for twin canted vertical tails on the T configuration. The vortex flow off the forebody and delta wing produced unstable pitching moments when it interacted with twin tails. Without horizontal tails, the aircraft did not have enough pitch authority to counteract these moments. A single vertical tail and no horizontal tails were finally identified as the best overall approach to the design despite the degradation of radar cross section in the side sector.

The Lockheed ATF design proposal featured highly swept trapezoidal wings and tail surfaces, twin canted vertical tails, and side fuselage-mounted engine inlets. The Lockheed design also incorporated a detachable rotary missile launcher in a central internal weapons bay. (The launcher was loaded away from the aircraft and then installed prior to

Fig. 21   Front quarter view of the General Dynamics ATF design.

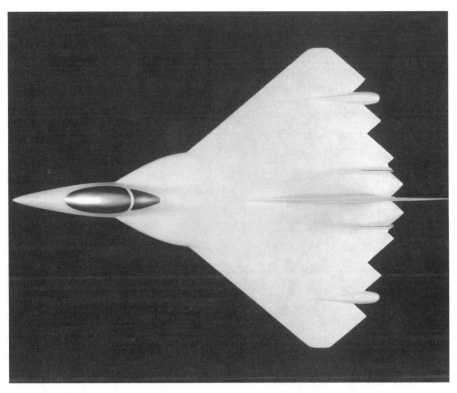

Fig. 22   Top view of the General Dynamics ATF design.

Fig. 23   Front view of the General Dynamics ATF design.

Fig. 24   Bottom view of the General Dynamics ATF design.

each mission.) It had three radar arrays, one in the nose and two located in the sides of the prominent fuselage forebody to achieve the 120-deg radar field of regard requirement on each side of the nose. Each wing-root incorporated an infrared search and track system that was designed to operate through faceted windows; these were intended to enable the aircraft to maintain its very low radar signature while still allowing operation of the infrared sensors. Photographs of a scale model of the Lockheed ATF Dem/Val design proposal are shown in Figs. 25–28.[86]

A presentation of the revised ATF program to the DSARC was planned for mid-June. This eventually slipped to mid-August. The review was finally held on August 19, 1986. This meeting was the main Milestone I decision review. Best and final offers (BAFOs) were requested from the bidders on September 15, 1986, and due on September 22. A follow-on meeting with the OSD principals was held on October 14 to address some specific issues prior to Dem/Val contract award. The Dem/Val source selection decision was announced, and contracts signed, on October 31, 1986 [although the official Milestone I decision memorandum was not signed by Mr. Godwin, undersecretary of Defense (Acquisition), until January 30, 1987].[16,87] A timeline for the entire concept definition phase of the ATF Program is presented in Fig. 29.

Fig. 25 Front quarter view of the Lockheed ATF Dem/Val design proposal.

Fig. 26 Top view of the Lockheed ATF Dem/Val design proposal.

Fig. 27   Front view of the Lockheed ATF Dem/Val design proposal.

Fig. 28   Bottom view of the Lockheed ATF Dem/Val design proposal.

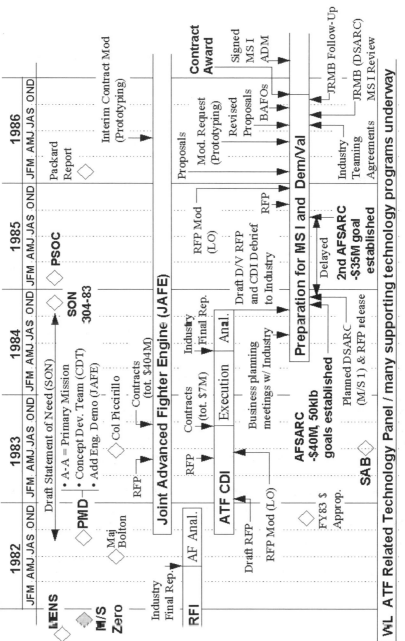

Fig. 29 Timeline—Phase 0 (concept definition).

REFERENCES

[1]Ferguson, P. C., *From Advanced Tactical Fighter (ATF) to F-22, Part I—To Milestone 0 and Beyond: 1970–1982,* Office of History, Aeronautical Systems Center, Wright–Patterson AFB, OH, May 1996.

[2]Sweetman, B., *YF-22 and YF-23 Advanced Tactical Fighters: Stealth, Speed and Agility for Air Superiority,* Motorbooks International, Osceola, WI, 1991.

[3]Twigg, J. L., *To Fly and Fight: Norms, Institutions, and Fighter Aircraft Procurement in the United States, Russia, and Japan,* Ph.D. Dissertation, Dept. of Political Science, MIT, Cambridge, MA, Sept. 1994.

[4]Piccirillo, A. C., Col. USAF (Ret.), Personal notes, 1983–1986.

[5]Patterson, R., Capt. USAF, *Wright Laboratory's Role/History in the Advanced Tactical Fighter Technology Development,* Wright Lab., Rept. No. WRDC-TM-90-603-TXT, Wright–Patterson AFB, OH, Jan. 1990.

[6]*White Paper on Advanced Fighters,* USAF/RDQT, with TAC/DRFG Comments, early 1982 (originally SECRET; declassified Jan. 10, 1997).

[7]Lyons, R. F., Maj. USAF, *The Search for an Advanced Fighter: A History from the XF-108 to the Advanced Tactical Fighter,* Air Command and Staff College, Air Univ., Maxwell AFB, AL, April 1986.

[8]Aronstein, D. C., and Piccirillo, A. C., *Have Blue and the F-117A: Evolution of the "Stealth Fighter,"* AIAA, Reston, VA, 1997.

[9]*Concept of Operation for the Advanced Tactical Fighter (ATF),* United States Air Force Tactical Air Command, Deputy Chief of Staff (Plans), Jan. 1971. SECRET. (Unclassified information only used from this source.)

[10]"New Soviet Aircraft Focus U.S. Attention on Advanced Fighter," *Aviation Week and Space Technology,* March 19, 1984, pp. 46, 47.

[11]*Mission Element Need Statement (MENS) for New Fighter Aircraft,* USAF/RDQ, first released July 1981.

[12]Green, W., *New Observer's Book of Aircraft,* Frederick Warne and Co., 1985.

[13]Message from HQ TAC to USAF/RDQ, Subject: *ATF Program Management Directive New Addition,* April 22, 1982. SECRET. (Unclassified information only used from this source.)

[14]Nordmeyer, Capt. USAF, TAC/DRFG, to TAC/CC, /CV, and /CS, Staff Summary Sheet: *Advanced Tactical Fighters (ATF),* May 17, 1982.

[15]Program Management Directive No. R-Q 7036(5) 63230F, *Advanced Tactical Fighter (ATF) Technologies,* USAF/RDQT, Aug. 24, 1982.

[16]Ferguson, P. C., *Advanced Tactical Fighter/F-22 Annotated Chronology* (Draft), Office of History, Aeronautical Systems Center, Wright–Patterson AFB, OH, Aug. 1996.

[17]Ferguson, P. C., *Oral History Interview: Brigadier General Claude M. Bolton, Jr.,* Office of History, Aeronautical Systems Center, Wright–Patterson AFB, OH, Aug. 13, 1996.

[18]*Semi-Annual History,* U.S. Air Force Systems Command, Aeronautical Systems Div., Directorate of Mission Analysis (ASD/XRM), July–Dec. 1982, p. 15.

[19]"ATF Is Delayed To Add Stealth Technologies," *Defense Week,* June 27, 1983, pp. 4, 5.

[20]Weekly Activity Reports (WARs), from ATF System Program Office to the Commander, Aeronautical Systems Div., July 13, Aug. 17, and Sept. 8, 1983.

[21]Haystead, J., "Sleek, Swift, Superior: A Look at Tomorrow's Advanced Tactical Fighter for the USAF," *Journal of Electronic Defense,* Nov. 1983, p. 53, 54.

[22]*Fact Sheet: Advanced Tactical Fighter,* U.S. Air Force Systems Command, Aeronautical Systems Div., Office of Public Affairs (ASD/PA), PAM No. 84-011, Aug. 1984.

[23]Meyer, D. G., "Futuristic Multirole Aircraft Could Guard High Frontier," *Armed Forces Journal International,* Sept. 1983, p. 48.

[24]"USAF Seeks Return to Fighter Traditions," *Flight International,* Oct. 8, 1983, pp. 938, 939.

[25]"Breaking Barriers to Build the Next Air Force Fighter," *Business Week,* Feb. 13, 1984, p. 128.

[26]Canan, J. W., "Toward the Totally Integrated Airplane," *Air Force Magazine,* Jan. 1984, pp. 34–41.

[27]Piccirillo, A. C., Col. USAF, Memorandum for Mr. Tremaine, Subject: *ATF Status Report,* July 28, 1983.

[28]ORD Attachment: *Requirements Correlation Matrix.* SECRET. (Unclassified information only used from this source.)

[29]Robinson, C. A., "USAF Reviews Progress of New Fighter," *Aviation Week and Space Technology,* Nov. 28, 1983, pp. 44–46, 51.

[30]Ulsamer, E., "Scoping the Technology Baseline," *Air Force Magazine,* Dec. 1982, pp. 24, 27.

[31]Banvard, K., "McClellan Will Maintain New Generation of Fighters," *Sacramento Union,* May 5, 1984, pp. A1, A2.

[32]Raimundo, J., "McClellan Will Fix, Maintain Stealth Jets," *Sacramento Bee,* May 5, 1984, pp. A1, A14.

[33]Hehs, E., "F-22 Design Evolution," Lockheed Martin Tactical Aircraft Systems, *Code One Magazine,* April 1998.

[34]Mullin, S. N., "The Evolution of the F-22 Advanced Tactical Fighter," Wright Brothers Lecture, AIAA Paper 92-4188, Washington, DC, Aug. 24, 1992.

[35]Rait, S., "Streamlining the Advanced Tactical Fighter," *Program Manager,* March–April 1985, pp. 2–5.

[36]"Growing ATF Program Transferred to Tactical Systems," *Air Force Times,* late 1984 [date unreadable].

[37]WARs, Oct. 31–Nov. 7 and Nov. 7–12, 1984.

[38]WAR, Nov. 7–12, 1984.

[39]WAR, Nov. 13–21, 1984.

[40]WAR, Nov. 28–Dec. 4, 1984.

[41]WAR, Dec. 26, 1984–Jan. 3, 1985.

[42]Various WARs, Feb.–April 1985.

[43]Preliminary System Operational Concept (PSOC) for the Advanced Tactical Fighter, HQ Tactical Air Command, Deputy Chief of Staff (Plans), Feb. 1, 1985.

[44]Bazley, Gen. USAF, Donnelly, Gen. USAF, and Kelley, Lt. Gen. USAF, Memorandum to Gen. Gabriel, May 14, 1985.

[45]Skantze, L. A., Gen. USAF, Letter to the Honorable Verne Orr, Secretary of the Air Force, May 10, 1985.

[46]"CBO Challenges ATF Assumptions," *Aerospace Daily,* April 30, 1985, p. 340.

[47]WARs, May 23–29 and Aug. 15–21, 1985.

[48]Orr, V., Secretary of the Air Force, Tasking Memorandum to Air Force Studies and Analyses, May 1985, quoted in WAR, May 29–June 5, 1985.

[49]WAR, May 29–June 5, 1985.

[50]WAR, June 13–19, 1985.

[51]WAR, July 17–24, 1985.

[52]WAR, July 3–10, 1985.

[53]WAR, Aug. 8–14, 1985.

[54]WAR, July 25–31, 1985.

[55]WAR, Aug. 22–28, 1985.

[56]WAR, Sept. 26–Oct. 2, 1985.

[57]Cooper, T. E., Assistant Secretary of the Air Force (Research, Development and Logistics), Memorandum to AF/RDCS, Subject: *Final Acquisition Action Approval for the Acquisition Plan for the Advanced Tactical Fighter (ATF) System,* Sept. 27, 1985.

[58]*Air Force Requests Proposals for Next Generation Fighter,* ATF press release, Oct. 8, 1985 [number illegible].

[59]Leuthauser, J., Maj. USAF/XRLA, *Talking Paper on Air Staff Proposals on ATF,* Dec. 5, 1985.

[60]Conver, S. K., Deputy Assistant Secretary of the Air Force (Programs and Budget), Memorandum to Mr. Aldridge, Subject: *ATF Reliability, Maintainability, and Producibility,* Nov. 15, 1985.

[61]WAR, Oct. 31–Nov. 6, 1985.

[62]Preliminary Senate Appropriations Committee (SAC) markup language on FY86 Air Force budge request, quoted in WAR, Oct. 24–30, 1985.

[63]WAR, Dec. 5–11, 1985.

[64]WAR, Jan. 30–Feb. 5, 1986.

[65]WAR, July 4, 1984.

[66]WAR, Jan. 16–22, 1986.

[67]WAR, Nov. 27–Dec. 4, 1985.

[68]WARs, Dec. 5–11 and 12–18, 1985.

[69]WAR, Feb. 13–19, 1986.

[70]"An Interim Report to the President," President's Blue Ribbon Commission on Defense Management, Feb. 28, 1986.

[71]WAR, April 3–9, 1986.

[72]WAR, May 1–7, 1986.

[73]WARs, April 3–9 and 24–30, May 1–7 and 8–14, 1986.

[74]Bedard, J., "Industry Raps ATF Plans," *Defense Week,* Nov. 3, 1986.

[75]WAR, March 20–26, 1986.

[76]WAR, July 8–15, 1986.

[77]WAR, May 22–28, 1986.

[78]Cooper, T. E., Assistant Secretary of the Air Force (Research, Development and Logistics), Memorandum for Vice Chief of Staff, Subject: *ATF AFSARC—Action Memorandum,* May 27, 1986.

[79]McMullen, T., Lt. Gen. USAF, ASD Commander, to ATF Dem/Val Offerors, Subject: *Advanced Tactical Fighter (ATF) Program Restructure,* May 28, 1986.

[80]WAR, May 29–June 4, 1986.

[81]WAR, June 12–18, 1986.

[82]Miller, J., *Lockheed's Skunk Works: The First Fifty Years,* Aerofax, Arlington, TX, 1993.

[83]WAR, July 24–30, 1986.

[84]Caption to Lockheed Martin Tactical Aircraft Systems photo P545-6101, undated.

[85]Caption to Lockheed Martin Tactical Aircraft Systems photo P545-6102, undated.

[86]Caption to Lockheed Martin Tactical Aircraft Systems photo P-545-6103, undated.

[87]Various WARs, Aug. 1986–Feb. 1987.

Chapter 4

PHASE I: DEMONSTRATION AND VALIDATION

OVERVIEW

Consistent with the early thrust of the ATF program, bidders for the ATF demonstration and validation (Dem/Val) phase were instructed to focus on risk reduction and technology development plans, rather than on specific aircraft designs, in their proposals. Even after the decision was made to fly ATF prototypes during Dem/Val, this overall strategy did not change. Every aspect of Dem/Val was oriented toward proving technologies and concepts, refining requirements, and reducing risk. The specific risk reduction activities and plans were formulated by the contractors, not the government. The Air Force specified goals for the Dem/Val phase, but the contractors were given the freedom to determine how best to achieve those goals.

The Dem/Val phase of the ATF program began with the award of the prime contracts to Lockheed and Northrop on October 31, 1986, valued at $691 million each. Based on prior industry agreements, these companies became the lead members of two teams: Lockheed/Boeing/General Dynamics and Northrop/McDonnell Douglas. In accordance with defense contracting practices that prevailed at the time, as well as economic conditions (many defense contractors competing for relatively few major programs), the Air Force was able to require firm fixed price (FFP) contracts for the Dem/Val phase and also was able to expect that the contractors would invest substantial sums of their own funding in order to stay competitive. "At the time the two firms were selected to compete, Program Manager [Col.] Piccirillo estimated that they would each spend from $300 to $500 million of their own funds on the competition."[1]

Of the $691 million value of each team's Dem/Val contract, approximately $100 million was allocated to radar and electro-optical sensors, $200 million for avionics architecture and integration, and the remainder for airframe and other miscellaneous tasks.[2] The engine companies were funded directly, and each would receive an additional $650 million during Dem/Val.

103

The Dem/Val phase of the ATF contract consisted of three principal activities—system specification development, avionics ground prototypes/flying laboratories, and prototype air vehicles, as described below:

1) *ATF system specification development (SSD):* Refine weapon system characteristics and operational requirements. SSD included radar cross section (RCS) tests, piloted simulations, material tests, reliability/maintainability demonstrations, and contributions to updates and refinements of the formal ATF acquisition requirements documents.
2) *Avionics ground prototypes/avionics flying laboratories (AGPs/ AFLs):* Demonstrate achievability of fully integrated avionics suites. Although not required to do so, both teams also chose to fly prototype avionics systems in airborne laboratories.
3) *Prototype air vehicles (PAVs):* Demonstrate the performance capabilities on which the operational ATF would be based. Each contractor (team) was to design, develop, fabricate, and flight test two air vehicles, one to be powered by General Electric engines and the other by Pratt and Whitney engines. The Lockheed prototypes were assigned the designation YF-22, with Northrop prototypes being designated YF-23.

The ATF Engine (ATFE) program became progressively more integrated with the total ATF program during this phase, culminating in the delivery and use of flight worthy YF119 and YF120 engines during the PAV flight-test program.

The initial plan was that the PAVs would make their first flights in late 1989. Full-scale development (FSD) proposals would be due in late summer of 1990, and the FSD source selection decision would be made by the end of 1990.[1] First flights by the two teams were actually made in August and September of 1990, and FSD [by that time renamed engineering & manufacturing development (EMD)] proposals were submitted in early January 1991. In April 1991, the announcement was made that the Lockheed/Boeing/General Dynamics F-22, powered by the Pratt and Whitney F119 engine, would proceed into EMD. The Milestone II Acquisition Decision Memorandum was signed on August 1, and the EMD contracts for the aircraft and the engines were awarded on August 2, 1991.

SYSTEM SPECIFICATION DEVELOPMENT

Throughout Dem/Val two distinct versions of each ATF design were maintained: the prototype air vehicle (PAV), which was the aircraft that each contractor was actually building during Dem/Val, and the preferred system concept (PSC), which was the aircraft that each contractor intended to build as the operational ATF. Although designs for the PAVs had to be finalized early in Dem/Val so that fabrication could proceed, evolution of the PSC designs continued. The PAVs were therefore not totally representative of each contractor's intended operational ATF, but rather were simply one part of the overall strategy to reduce risks and refine requirements during Dem/Val. PAV flight test experience, PSC design evolution, and all of the other demonstration activities were closely interrelated with the development of ATF requirements and definition of the system specification.

INITIAL SYSTEM SPECIFICATION

System requirements reviews (SRRs) were held from April 27–30, 1987 (at Northrop) and May 4–7 (at Lockheed). The contractors presented results of performance and cost trade studies for the Air Force to use as the basis for decisions on ATF requirements. This process was intended to lead to a draft system specification early in Dem/Val. The specification would then be revised and refined as design, analysis, and demonstration results provided additional insight into what was technically feasible, as well as in response to ongoing definition of the threats.

Data presented at the SRRs were analyzed by the Air Force from May to August 1987. The Air Force evaluation of the contractor trade studies was briefed to the contractors in late June, and subsequently to TAC, the Air Staff, etc. In August, HQ TAC (Headquarters for Tactical Air Command) informed HQ AFSC (Headquarters for Air Force Systems Command) of concern that the PAVs might not be close representations of the PSCs and that the gross weight of the PSCs could be as high as 60,000 lb, which in turn led to an estimated unit flyaway cost of around $40 million (vs the "official" goals of 50,000 lb and $35 million, although $40 million was always a much more realistic estimate). At the formal headquarters-level presentation of the SRR results to TAC on August 19, Gen. Robert Russ (TAC/CC) "asked that the SPO and TAC look for the best combination of trades at this time to reduce the PSC weight to 50,000 lb." The SPO worked with TAC to investigate requirements changes that would be necessary to attain the program weight and cost goals.

Proposed trades were briefed to the TAC leadership, the Air Staff, the Air Force Council, and the secretary of the Air Force in September and October. Col. Fain (who became program director in December

1986 following Col. Piccirillo's retirement) made many presentations during that time, and "Brig. Gen. [Joseph] Ralston [HQ TAC/DR] described the rationale behind the proposed requirements trades. He also provided a requirements correlation matrix between the TAC System Operational Requirements Document (SORD) and the [November 1984] Statement of Operational Need (SON)...."[3,4]

Some adjustments were made to the air vehicle performance requirements during that period. A specified cruise altitude was eliminated from certain mission profile descriptions, allowing the optimum cruise altitude to be used instead. Certain maneuver requirements that appeared to be driving the design to higher gross weights were relaxed somewhat.[1]

Another change was made during 1987 as a result of F-15 STOL/maneuver technology demonstrator (S/MTD) ground test results.[5,6] The S/MTD had experienced difficulties in the development of its two-dimensional vectoring/reversing nozzles, including weight growth and cooling problems. The vectoring requirements in that program were reduced slightly to allow the nozzles to be built for a tolerable weight penalty.[7] Even so, the weight of the S/MTD nozzles was higher than what would be desired for a production fighter. The cooling required during thrust reverser operation was identified as a major weight driver. These experiences led the Air Force to delete the requirement for thrust reversing for the ATF and, as a consequence of that decision, to relax the STOL requirement from 2000 to 3000 ft. "Even if the entire F-15 S/MTD program was to accomplish nothing else—which was not the case," the total cost of the S/MTD program "would have been a small fraction of the cost of discovering the same problems in FSD [of the ATF]."[5]

The first draft of the preliminary system specification (PSS) was published in November 1987. The draft system operational requirements document (SORD) was released in December 1987. Compared to the November 1984 statement of operational need (SON), the 1987 SORD contained several new or refined requirements commensurate with the increased level of definition in the program. The SORD also contained the changes described above to reduce risk and bring the ATF closer to "50,000 lb and $35 million."[4,8] However, the cost and weight issues were by no means resolved at that time. Cost and weight control continued to be critical issues in the ATF program throughout Dem/Val and in the EMD phase through the present time.

Up to this time the ATF engines were being developed to produce approximately 30,000-lb thrust based on an assumed ATF takeoff gross weight of 50,000 lb. Despite intense efforts by both Northrop and Lockheed focused on achieving a 50,000-lb aircraft, the engine companies were informed in late 1987 that the production engines would need to increase in size to the 35,000-lb thrust class. Aircraft weight and drag expected of the ATF contractor designs (the large internal fuel volume necessary to meet ATF combat radius requirements was an important contributor) were sufficiently larger than when the initial engine designs were formulated, so that both General Electric and Pratt and Whitney would have to make significant changes in their designs in order for the ATF aircraft to be capable of meeting supercruise objectives.

REQUIREMENTS EVOLUTION, 1987–1989

During and after the development of the preliminary system specification, the Lockheed/Boeing/General Dynamics team experienced difficulties arriving at an acceptable design. According to the team's General Manager, Sherman Mullin:

> We faced two major areas of issues during the ATF Dem/Val program:
> 1) System engineering issues, based on performance requirements versus flyaway cost and weight. Many initial "require ments" which seemed to be no problem turned out to be major drivers of weight and/or cost, and, as a result, were substantially changed by the Air Force, based on our trade study data.
> 2) Design engineering issues...such as supersonic drag versus internal fuel and weapons load, supersonic and subsonic maneuverability versus low observability, installed propulsion system performance versus low observability, and many others.
> Over the period from late 1986 to the fall of 1988, the progress in resolving these issues was slow. Hope sprung eternal that a 50,000-lb TOGW airplane was achievable....[9]

However, that hope was never realized, and it became increasingly evident that further revisions to the requirements were necessary:

> In late 1988 we got on the road to success. The appropriate USAF generals got involved, and based on trade study data, adjusted the requirements to achieve the acceptable balance of system performance and cost, with weight as a surrogate for cost.... Initially most of our engineers had a full scale development mentality and treated stated requirements as fixed. They were more in-

terested in designing an airplane than performing numerous trade
studies. Then we really started to do some solid studies of technical
performance versus weight and/or flyaway cost. Initially the Air
Force had difficulties making decisions based on trade study data,
but by late 1988 they were successfully making decisions. By the
time we completed the formal System Design Review in November 1989 it was clear that the convergence of ATF performance,
weight, and cost was being achieved and that the specific system
engineering approach we had implemented was really working. We
never did find the magical 50,000 lb, $35 million airplane, but we
found the "right" answer.[9]

By 1988–1989, it was finally beginning to be acknowledged that
ATF takeoff gross weight would probably be somewhat higher than
50,000 lb. However, every effort was made to ensure that the ATF
would be affordable while maintaining its essential mission capabilities.
Air vehicle performance requirements were adjusted to relax certain
points that were significant size/weight drivers but were of minimal operational value. Some avionics capabilities were also revised to reduce
risk and/or cost. The requirement for radar side arrays was deleted.
Software spare/growth requirements were temporarily removed pending a better definition of the need. The infrared search and track
(IRST) system was changed from a threshold (i.e., requirement) to an
objective (i.e., goal).
    The requirement for the IRST had already gone through several
evolutions. The original requirement, in the November 1984 ATF
statement of operational need, was for a multicolored IRST. In 1987,
the requirement was changed to single-color, with multicolor as a
goal. In 1989, the inclusion of any IRST was downgraded to a goal to
reduce cost, weight, and technical risk. Nevertheless, IRST systems
were tested in the avionics flying laboratories. In 1990, a requirement
was added to have growth provision for an IRST (including physical
space/power/cooling provisions and avionics processing provisions to
allow cross-cueing and fusion of IRST data on cockpit displays). This
requirement would reduce the cost, weight penalty, and risk of adding
one later on.[10]
    Other miscellaneous capabilities that had been included in the 1984
SON, and carried over in the 1987 SORD, were looked at more critically during the 1998–1999 revision period. For example, the nuclear
hardening requirement was deleted at this time. TAC agreed to reduce
the ejection seat capability from "ejection throughout aircraft flight envelope" to ACES II capability to allow use of the ACES II seat in the
ATF for FSD and production, with two provisos[10,11]:

1) AFSC should continue to pursue advanced seat development outside of the ATF program.
2) The ATF should at least retain growth potential for an expanded seat capability.

Preparatory to the release of a revised version of the SORD, a 1988 "Fall Requirements Review" was conducted, involving AFSC and TAC. This review addressed possible ATF alternatives (F-15/F-16 derivatives) and included a reassessment of the threat. Following this review a revision to the SORD was released in February 1989.

GUN QUESTION

One issue that received a lot of attention in the press at that time was whether the ATF would carry a gun. This issue was discussed and debated repeatedly from program initiation through Dem/Val and into EMD. Lt. Gen. Thomas H. McMullen, commander of the Air Force Systems Command's Aeronautical Systems Division (ASD), was quoted in early 1984 as stating that an air-to-air gun "probably will be installed" on the ATF, but that further trade studies would be conducted.[12] Another unnamed source commented in a letter to *Aviation Week and Space Technology,* "I sincerely hope that Col. Albert C. Piccirillo is serious in his opinion that a gun will be included in the advanced tactical fighter (ATF) armament package.... When the missiles fail, are spent or are effectively countered," having a gun provides "graceful system degradation."[13] The operational community was unwilling to accept the idea of an air superiority fighter without a gun, and the formal requirements documentation consistently included a gun as part of the ATF's armament.

Given that there would be a gun, the next question was what kind of a gun. Earlier debate (early 1980s) had centered around what caliber gun should be carried. By the time Dem/Val started, with the role of the ATF solidly defined as air-to-air, it was generally agreed that 20 mm or equivalent was optimum for the ATF's mission. There remained the choice between the proven M61 cannon, or a new gun such as might emerge from the advanced gun technology (AGT) program. A flightweight AGT gun was successfully tested in 1989, although significant development remained to be accomplished before such a gun could be considered ready for production. Proponents of the new gun technology pointed out that the M61 was an old design, that new technologies such as telescoped ammunition offered about a 50% increase in velocity together with a decrease in volume and weight of the ammunition, and that the development costs would be insignificant compared to that

110 ARONSTEIN, HIRSCHBERG, AND PICCIRILLO

of very high-speed integrated circuit (VHSIC) avionics, new engines, and other ATF technologies and systems.[13]

However, the cost impact of a new gun would not be limited to development. A new gun that was unique to the ATF could be expected to cost more in production, and most importantly the logistical impacts would be considerable if a new gun with unique ammunition were used on the ATF. At the same time, the improvement offered by the new gun (primarily an increase in muzzle velocity from 3600 to 5000 fps) is important primarily at relatively long range where missiles become the weapon of choice anyway. As a result of these factors, the acquisition of proven gun technology became the preferred choice. The M61A2 was therefore selected as the gun for the ATF with a lengthened barrel and other moderate improvements.[5]

LATE DEMONSTRATION AND VALIDATION REQUIREMENTS UPDATES

The next major milestone in ATF requirements development was the system design review (SDR) in late 1989. This was especially significant in the area of avionics, since the SDR came only a year after the commitment to integrated modular avionics architecture and was therefore an important opportunity to define avionics performance requirements and functional capabilities within the overall integrated architecture.

The SORD was updated again in April 1990 and finally in March 1991 (corresponding to Milestone II, the end of Dem/Val, and the beginning of EMD). The 1990 update began to define pilot training system requirements, commensurate with the need to plan this aspect of the EMD phase, along with the increased emphasis being placed on training in the acquisition process. Both the 1990 and 1991 revisions incorporated information that was gathered from the many demonstrations that took place, particularly during the later stages of Dem/Val. The demonstrations provided increased confidence that some requirements could be met, while revealing other areas in which the original requirements were not technically feasible, or carried excessive risk.

Overall, the Dem/Val phase was characterized by a flexible approach to requirements. This is consistent with the overall purpose of a Dem/Val phase, which is to produce a convergence between what is required and what is technically feasible. In 1990 the General Accounting Office (GAO) reported that the approved program baseline "contained only technical characteristic 'goals' for the system," and that specific characteristics and performance thresholds would not be set until the system specification for FSD was written at the completion of the Dem/Val phase.[14] This reflected an important part of the purpose of the

Dem/Val phase: to reduce risk and to provide information on what capabilities could be achieved. Setting too many hard requirements prior to the completion of the Dem/Val testing, and in particular the prototype flight tests, would have defeated that purpose. However, this strategy was not universally accepted or understood. In 1990, the GAO reported:

> As of February 1990...some critical cost and performance trade-off decisions and system demonstrations still need to be completed. The ATF's design and system specifications are subject to change, and the Air Force will not assess the ATF's performance capabilities for FSD until the required cost and trade-off studies, engineering analyses, component tests, and prototype demonstrations have been completed. Nevertheless, the Air Force has been directed by the Defense Acquisition Board [DAB] to submit the specific radar cross section, supersonic cruise, maneuverability, mission radius, and integrated avionics capabilities to be achieved during FSD before the evaluations are completed.[14]

This predicament illustrates the conflicting pressures that any acquisition program may face. Program managers are expected to take responsible actions to control cost and, at the same time, are expected to commit to specific capabilities that the weapon system will provide. However, in the interest of cost control it is highly desirable to leave certain options open until cost/performance tradeoffs are better defined. Micromanagement of detailed weapon system characteristics at too high a level may restrict the program manager's ability to develop a balanced, affordable weapon system.

AVIONICS GROUND PROTOTYPES AND FLYING LABORATORIES

A significant part of the early avionics effort (first six months) was devoted to scrubbing the requirements in conjunction with the overall ATF system requirements review (SRR) process. As in every area of the ATF's requirements, the objective in avionics was to eliminate requirements that were of marginal value but high cost or risk. Following the SRR, the focus shifted to demonstrations. Major avionics demonstration objectives included the following[15]:

1) Identify appropriate advanced technologies that were relevant to meeting ATF requirements.
2) Reduce risk through hardware and software demonstration. Demonstrate key capabilities, both of the overall integrated ar-

chitecture and of specific functional capabilities such as sensor performance.

3) Demonstrate the use and suitability of the Ada programming language for all avionics software. Demonstrate mature, production quality software development tools. Put the necessary procedural arrangements, interface control groups (ICGs), and other aspects of teaming/subcontracting together during Dem/Val so that the required structure would be in place at the beginning of FSD.

4) Refine and document avionics requirements. The demonstrations were all highly interrelated with system specification development. Data from the demonstrations were used to progressively refine the requirements and to provide additional levels of detail in areas such as required processor performance (throughputs and memory capacities), interconnect performance (data rates and latencies), sensor and display requirements, and the appropriate level of information correlation and decision support to be provided to the pilot.

5) Demonstrate features for improved reliability, maintainability, and supportability (RM&S), including modular hardware packaging, liquid-cooled module enclosures, embedded interconnects, and built-in fault detection. These features, together with the overall integrated architecture and software advances, were intended to double the reliability while cutting maintenance requirements in half, relative to then-current fighter aircraft avionics.

The ATF Dem/Val efforts in avionics included two parallel areas: 1) system architecture and core processing and 2) individual functional capabilities. The overall avionics suite is shown schematically in Fig. 30. The core architecture effort focused on developing and demonstrating an integrated modular avionics architecture with the associated programming and processing capabilities. In parallel, other efforts within the ATF program and in separate supporting technology projects concentrated on achieving improvements in performance, weight/cost, RM&S, or integration of specific functions or sets of functions such as the radar, CNI (communication, navigation, and identification) systems, cockpit displays, etc. Some of the efforts that took place in separate projects are discussed in Chapter 5.

The first task for both ATF teams was to define the work breakdown among the members, establish relationships with subcontractors, etc., to ensure that all necessary elements would be in place building up to the major ground and flight demonstrations. Demonstrations started with

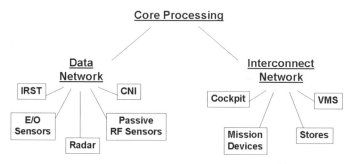

Fig. 30   Schematic of ATF avionics suite.[16]

laboratory breadboard/brassboard tests, many of which were already ongoing under other programs. Then the AGP demonstrations began, starting with the core processing prototypes and adding progressively more functions and interfaces, and finally tests were conducted using the avionics flying laboratories (AFLs). Although flying avionics prototypes were not required in the RFP, both of the winning contractors chose to include AFLs in their proposed Dem/Val efforts. The flying laboratories were therefore included in the Dem/Val contracts that were awarded in October 1986. The Lockheed/Boeing/General Dynamics team planned to mount a prototype ATF avionics suite in the prototype 757 commercial transport, which Boeing maintained and operated as a flying testbed for various purposes. This aircraft is illustrated in Fig. 31.[2] Separately, the Northrop team proposed to use a

Fig. 31   The Boeing 757 avionics flying laboratory was used in support of the Lockheed team's ATF Dem/Val program.

Fig. 32   The Northrop team used a modified BAC 111 airliner as their avionics flying laboratory.

modified British Aerospace Corporation BAC 111 airliner as the platform for their AFL. This aircraft is seen later in the ATF Dem/Val program in Fig. 32.[17]

Like the prototype air vehicles (PAVs), the avionics prototypes were not intended to achieve the full projected ATF avionics capabilities, but rather to verify that the projected performance was achievable and that the technology was mature and available. All of the demonstration activities took place in conjunction with analysis and simulation and were closely interrelated with the requirements development process. Col. Michael Borky, the avionics manager at the System Program Office (SPO) in 1990, reported:

> This analytical work has drawn upon the data produced in avionics prototype testing and, conversely, has helped to guide the evolution of the prototypes and the process of setting the avionics baseline for the start of FSD. In addition, each successively refined generation of the YF-22 and YF-23 avionics Preferred System Concepts (PSCs) has been validated through man-in-the-loop simulation with experienced fighter pilots to ensure that the cockpit and avionics functions provide what the aircrew needs and expects. At every stage, the evolution of the avionics baseline has been fully coordinated with the Tactical Air Forces (TAF) to ensure the proper priorities and functional requirements are applied.[15]

Lockheed's first AGP demonstration took place in October 1988.[18] Northrop also began to conduct AGP demonstrations around that time; they eventually would demonstrate pilot-controlled real-time fusion of multisensor flight data, the integration of 600,000 lines of Ada software code (developed using common software tools), a core processing capability 100 times faster than that of then-current air superiority fighters, a fully integrated advanced avionics architecture, self-diagnostics/fault isolation, and system reconfiguration.[18] As Col. Borky noted,

> By about the midpoint of the Dem/Val phase, both contractors had assembled and exercised their initial core processing proto-types. This was essential for several reasons. First, it established the processor characteristics and interfaces against which sensor, EC [electronic combat], CNI [communication/navigation/identifica-tion], and other subcontractors could design and integrate their parts of the avionics suite. Second, it provided the data which al-lowed the program to commit firmly to the modular, integrated concept. And finally, it allowed the contractors to refine their de-signs and firm up core processing requirements for FSD. This set of demonstrations also provided extensive experience with Ada pro-gramming in avionics applications.[15]

Thereafter, functional prototype results were progressively integrated with the core prototypes. The full avionics ground prototypes included the core prototypes with prototype processors and Ada software, con-nected to a controls and displays bench, environment/scenario simula-tions, and real-time data inputs from the prototype sensors and other functional avionics subsystems.

The AFL demonstrations then provided improved fidelity for evalu-ating clutter effects on sensors and similar issues. For example, in-flight tests with the radar demonstrated detection ranges equal to or greater than calculated performance against fighter-sized targets flying typical engagement profiles. The favorable flight results supplemented ground test data that validated the beam shaping capability, low side lobes, phase accuracy across the array, fault tolerance, reliability, power and cooling parameters, and other critical aspects of active array technol-ogy. In total, the demonstrations showed that active array technology was thoroughly understood and did not pose a significant technical risk for FSD. Integration of prototype radars with core processing further validated the feasibility of the selected radar operating modes and mode interleaving and provided sound estimates of software and pro-cessing requirements.[15]

Each team completed approximately 100 flight hours with their re-spective avionics testbeds during 1990. These tests validated the ability to "detect a target by multiple sensors and display it reliably and consis-tently as one symbol on the pilot's display."[5] The Lockheed team's AFL, the Boeing 757, began flight tests on April 18, 1990, and flew for approx-imately four months. Sensors tested aboard the Boeing 757 AFL in-cluded the Texas Instruments/Westinghouse active array radar, the TRW

CNI system, the Lockheed Sanders/General Electric electronic combat system, and the General Electric infrared search and track (IRST) system. The Northrop team, using their BAC 111 AFL, demonstrated an active electronically scanned antenna (ESA), all-aspect threat missile launch and detection and tracking capability, and an imaging IRST system. The installed performance of each sensor, the integrated avionics suite, and the mission avionics sensor management and sensor track integration functions were evaluated. Several tests were conducted against targets of opportunity, including commercial, military, and general aviation aircraft.[19] Col. Borky summarized the contributions of the AFLs as follows:

> Although...the details of the two contractor teams' approaches to the use of the AFLs were different, both have conducted demonstrations which yield the desired evaluation of sensors and other avionics elements to complete the data base upon which the FSD program will be founded. Moreover, this testing has shown the great power of a large, relatively benign aircraft environment as a system integration facility intermediate between the ground laboratory and the actual fighter aircraft. This concept of an airborne test bed is a central feature of the ATF FSD strategy.[15]

The above discussion illustrates how contractor innovation can favorably impact the conduct of a test program (remembering that the AFLs were not a required part of Dem/Val, but were independently proposed by the contractors).

The Dem/Val avionics ground and flying prototype efforts' collective demonstration accomplishments are summarized in Fig. 33.

	RANGE	MODES/FUNCTIONS	WAVEFORMS	THROUGHPUT	MEMORY	TIMING/LATENCY	BUILT-IN TEST	FAULT RECOVERY
RADAR SENSOR	X	X	N/A	X	X	N/A	X	X
IR SEARCH & TRACK	X	X	N/A	X	X	N/A	X	X
ELECTRONIC COMBAT	X	X	X	X	X	N/A	X	X
COMM/NAV/ID	X	X	X	X	X	N/A	X	X
CORE PROCESSING	N/A	N/A	N/A	X	X	X	X	X
SYSTEM INTERCONNECT	N/A	N/A	N/A	X	N/A	X	X	X
CONTROLS & DISPLAYS	N/A	X	N/A	X	X	X	N/A	X

Fig. 33   Dem/Val avionics demonstration matrix.[16]

SOFTWARE

The ATF avionics prototypes represented "one of the largest Ada developments ever undertaken" up to that time. The suitability of Ada as the programming language for the avionics suite on a fighter aircraft was demonstrated. Furthermore, software engineering tools were selected, matured, integrated, and demonstrated, paving the way for FSD. A series of software development capability and capacity reviews were conducted for all contractors and subcontractors with major software responsibilities. "The bottom line is that Ada is now proven and overwhelmingly preferred for virtually all software development in the ATF and other weapon system programs."[15]

LESSONS LEARNED FROM THE DEMONSTRATION AND VALIDATION AVIONICS GROUND PROTOTYPE/AVIONICS FLYING LABORATORY

Technology transition from laboratory to weapon system programs demands a two-way flow of information. The companies performing the laboratory programs must know what the real technology needs are for the weapon system, and the companies developing the weapon system must be kept fully informed on the status of the technology. In the ATF program, "Contractually required Associate Contractor Agreements (ACAs) between the ATF and technology base program contractors, along with access to design and test data and continuing management emphasis, helped ensure that the necessary exchanges of information took place."[15]

"Use of advanced technologies requires constant attention to risk management. Flexibility to employ only those technologies which produce benefits in system performance, cost, or supportability must be maintained. Every technical risk area must have one or more less aggressive backup approaches identified along with decision milestones and criteria for choosing between the primary and fallback designs."[15] A good example was the commitment to integrated, modular architecture in late 1988. The program was structured so that this decision would occur after the first core prototype demonstrations had validated the basic concept and architecture, but still early enough to support the remainder of the avionics program (i.e., the incorporation of the various functional avionics capabilities and systems with the core processing in accordance with the architecture chosen at the decision point).

Strong early attention to program planning and coordination "is essential to ensure that analysis, manned simulation, and prototyping activities support each other and the overall weapon system development."[15]

"Effective working relationships between the SPO and the contractors are important and were emphasized from the very start of Dem/Val," producing a team rather than adversarial relationship.[15]

Prototype Air Vehicles Development and Flight-Test

Work Breakdown

Distribution of work among the teaming partners was the first challenge. For the Lockheed team, "The overall distribution of work between the team partners was based on the dollar value of work performed.... In simplest terms, the program was divided into thirds, which in actuality was no simple task considering the interfaces that were required by this arrangement."[20] Because the major tasks associated with designing and building a weapons system do not logically fall into perfect thirds, it was necessary to further subdivide some of the tasks. The eventual distribution of responsibilities was as follows:

1) *Lockheed*—Weapon system, air vehicle, and avionics system design integration, forward fuselage, flying surface leading edges and tips, and final assembly.
2) *Boeing*—Wing, aft fuselage, propulsion system integration.
3) *General Dynamics*—Mid-fuselage, tail, most subsystems, armament system, landing gear, vehicle management system (including flight controls).

In addition, there were approximately 650 subcontractors in 32 states supporting the Lockheed/Boeing/General Dynamics team in the YF-22 PAV effort and associated ATF Dem/Val activities.

On the Northrop/McDonnell Douglas team, Northrop was responsible for most of the design and engineering effort, total systems integration, final assembly, construction of the aft fuselage and tail surfaces, and flight control system integration. McDonnell Douglas was responsible for the forward and center fuselage sections, landing gear, wings, fuel and armament systems, controls, and cockpit displays. The two companies shared responsibility for the crew station and pilot-vehicle interface. In avionics (although this was not part of the PAV effort), Northrop had primary responsibility in the area of defensive avionics, while McDonnell Douglas was the lead for offensive avionics.[19]

Early Demonstration and Validation Design Effort

The original Dem/Val schedule called for the ATF prototype designs to be "frozen" during mid-1987.[2] However, for the Lockheed team, "By the middle of July 1987, the existing aircraft configuration was determined by the team to be technically and competitively unacceptable, and new configuration design and development work was initiated on July 13."[21] After a very intense three-month effort, a new configuration was selected in October 1987, which, after further iterations of refinement and detailed optimization, became the final YF-22A design. Among the changes were the diamond-like wing and tail planforms (features which had been part of General Dynamics's design) replacing Lockheed's original, slightly less-tapered trapezoidal surfaces, and a significant reduction in the planform area of the forebody to improve the high angle-of-attack pitch characteristics.[19] In April 1988, the prototype configuration was temporarily "unfrozen" to reduce supersonic drag. This was accomplished through a redesign of the forward and aft fuselage sections in May.[9]

Northrop, on the other hand, had already settled on their aircraft design concept even prior to the decision to build prototypes.[22] They did not make any dramatic changes, and the basic YF-23 design was apparently finalized as planned during 1987. The Air Force operational requirement for ATF thrust reversing capability was subsequently deleted. Rather than redesigning the aircraft, the YF-23 team reportedly built the two prototype aircraft with somewhat oversized engine nacelles that were designed to house the originally intended thrust reverser systems.[5]

Emphasis shifted to detail design during 1987–1988, followed by fabrication of both teams' prototypes. Final assembly of the aircraft began in late 1989 or early 1990. Concurrently, both teams continued to evolve their PSC designs in response to the ongoing requirements adjustments, various ground demonstrations, and experience gained in the design, analysis, and testing of the PAVs.

THRUST VECTORING

The Lockheed team chose to use thrust vectoring on their ATF, whereas the Northrop team did not. Thrust vectoring as used on the YF-22 and F-22 provides the ability to attain and trim at very high angles of attack and increased pitch rates both in low speed and in supersonic flight. Furthermore, although the nozzles do not vector differentially, the use of vectoring increases the roll response and maximum roll rate attainable at high angle of attack. This is because the horizontal tail surfaces (stabilators) perform double duty, providing both pitch and (in conjunction with the ailerons and flaperons) roll control. The use of thrust vectoring for additional pitch control leaves more of the stabilator motion available for roll.[19] Vectoring capability reportedly only added

30–50 lb to each nozzle of the YF-22,[9] although there was an additional weight penalty on the structure that supported the vectoring loads.

An important question in the use of thrust vectoring is how much capability must be retained in the aerodynamic control surfaces. Conceptually, thrust vectoring can be used to replace and/or reduce the size of aerodynamic controls. This could lead to reduced aircraft weight, size, and cost. However, control would be degraded (or even lost) if the engine or the vectoring mechanism failed. The other approach is to require that the aircraft be fully controllable with aerodynamic controls only, so that loss of power and/or a malfunction of the vectoring nozzle does not (in theory) lead to loss of control. In this case the aerodynamic controls have to be sized approximately the same as if there were no vectoring. Vectoring in this case can still provide added capability, as it does on the F-22, but cannot be used in any significantly cost-reducing way.

The YF-22 was designed to be fully controllable as well as capable of recovery from any angle of attack within its flight envelope without the use of thrust vectoring. Thus the YF-22's tail surfaces were not appreciably smaller than they would have been if Lockheed had elected not to use vectoring at all. However, thrust vectoring gave the YF-22 much greater agility under certain conditions than the aircraft would otherwise have had. This enhanced capability is illustrated in Fig. 34; it de-

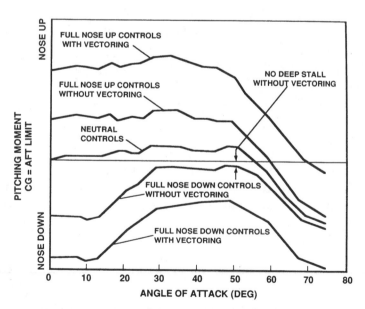

Fig. 34 The YF-22's use of thrust vectoring provided a major increase in pitch pointing capability over a wide range of angle of attack.[23]

picts the great increase in available YF-22 pitching moment (with corresponding improvements in aircraft stability and controllability) that were made available through the use of pitch-axis thrust vectoring.[23]

WIND-TUNNEL TESTING

The scope of the ATF Dem/Val wind-tunnel test program was unprecedented both in its sheer magnitude and in the quantity of data that were collected for a program in this early phase of development. Development of each of the competing ATF prototype designs consumed a vast amount of wind-tunnel test time. During these tests, a wide variety of industry and government facilities were used. Even prior to Dem/Val contract award, Lockheed reportedly used approximately 7000 h of wind-tunnel test time. A further 18,000 h of wind-tunnel testing was conducted by the Lockheed team during the "formal" Dem/Val contract phase. Because of the importance attached to supersonic cruise performance and high maneuverability in the ATF operational requirement, much of the Lockheed ATF team's wind-tunnel test focus was on refining the design to reduce overall drag and optimizing the stability and control aspects of the YF-22 design. The YF-22's use of thrust vectoring, as well the implications of high angle-of-attack (AOA) maneuvering on engine/inlet compatibility, resulted in a major emphasis on propulsion aspects in the Lockheed team's wind-tunnel test program. Edsel R. Glasgow of Lockheed, chief flight sciences engineer for the Lockheed ATF team, had overall responsibility for YF-22/F-22 wind-tunnel testing.

The specific types of wind-tunnel tests conducted by the Lockheed ATF team during the YF-22 Dem/Val effort, along with the number of hours dedicated to each category, are listed in Table 1.[23]

Table 1 Lockheed ATF team wind-tunnel tests

Tests conducted	Hours dedicated
Low-speed stability and control	6715
Propulsion aerodynamics	5405
High-speed drag and stability and control	4000
Weapons environment and separation	700
Flow visualization	695
Air data	240
Pressure model (airloads)	95
Flutter	80

In addition to its use in evaluation of various YF-22 aircraft design configurations and the determination of associated performance and structural loads, the aerodynamic database that resulted from this intensive wind-tunnel effort was essential in defining the design of the YF-22 handling qualities simulator. This aerodynamic database contained more than 730,000 individual data points; it incorporated six-degree-of-freedom aerodynamics, forced oscillation and rotary balance data as well as engine-induced jet effects. In obtaining high AOA data on the YF-22 alone, nine different models were employed in an intensive series of tests conducted at a variety of government and industry facilities. Wind-tunnel models and their roles in the YF-22 wind-tunnel test program are summarized in what follows.[23]

1) *1/20-Scale stability and control model:* Sting-supported, this model was used to measure forces and moments in all three axes. The model was designed to enable evaluation of changes in aircraft configuration during the detail design process. Flight surfaces (wings/vertical tails) could be moved and/or replaced. All control surfaces (leading-edge flaps, ailerons, flaperons, rudders, and horizontal tails) could be deflected. Landing gear, gear doors, weapons bay doors, speedbrake, and deflected nozzle flaps could be externally mounted on the model. Engine nacelles were faired over. Testing was conducted in the Lockheed 8 × 12-ft low-speed wind tunnel.[23]

2) *1/25-Scale stability and control model:* Used to provide primary stability and control data across the entire YF-22 flight envelope and equipped with an internal force balance. Flow-through inlets were accurately modeled, and internal flows were measured. All control surfaces were manually movable to actual aircraft limits. Testing was conducted in the NASA Langley 4-ft supersonic wind tunnel, the Calspan 8-ft transonic wind tunnel, and the General Dynamics 8 × 12-ft low-speed wind tunnel.[23]

3) *1/7-Scale free flight model:* This high fidelity, dynamically scaled model was equipped with electro-pneumatically operated actuators that drove all control surfaces with the exception of the leading-edge flaps, which were manually deflected. The model was used to evaluate both aerodynamics and high AOA flight control laws. Compressed air ejectors that entrained inlet flow through each duct and exhausted out of the vectoring nozzles were used to provide thrust. The model even included such details as the flight-test boom and simulated air data probes. Test-

ing was conducted in the NASA Langley 30 × 60-ft low-speed wind tunnel. Three pilots located in the wind-tunnel control room individually maintained pitch control, roll/yaw control, and thrust control. Model maneuvering rates and attitudes up to an AOA of 59 deg were measured during the free flight model test program. This model was also used for static and forced oscillation tests while mounted on a support system in the same wind tunnel.[23]

4) *1/14-Scale high AOA rotary balance model:* All control surfaces on this model were manually deflectable. Tested in the NASA Langley spin tunnel, the 1/14-scale model was mounted on a special support system that rotated the model about the relative wind vector. Tests using this model provided aerodynamic derivatives as a function of rate of rotation. This enabled aircraft spin modes to be calculated and configuration-specific characteristics such as nose slice and wing rock to be determined. Data developed from tests on this model were used in high AOA computer simulations during the definition of the YF-22's flight control laws. These simulations were also used by Lockheed test pilots during evaluation of YF-22 high AOA characteristics including spin resistance, departure from controlled flight, and spin recovery.[23]

5) *1/30-Scale free spin model:* This small, hand-launched dynamically scaled model was used to support definition of YF-22 spin modes and determination of recovery characteristics. Spin modes and recovery techniques predicted by testing using the 1/14-scale rotary balance model were validated with this model. All control surfaces on the model were remotely actuated to effect spin recovery. Testing was accomplished using the NASA Langley vertical spin tunnel.[23]

6) *1/10-Scale inlet/forebody compatibility model:* Used to provide initial inlet data on total pressure recovery, steady-state distortion, turbulence, and inlet instability. The model was later used to support development of forebody shapes (eventually leading to the YF-22 configuration). It also helped to define the location for the forebody-mounted air data probes and to develop inlet bleed air and bypass geometry. Testing was accomplished in the Lockheed 4-ft trisonic wind tunnel and the Boeing 9-ft low-speed wind tunnel.[23]

7) *1/7-Scale inlet/forebody compatibility model:* This very high fidelity verification model was used to simulate the final YF-22 configuration; it was highly instrumented and was coupled to a

computer system that enabled real-time screening of instanta-
neous engine front face distortion patterns. Results from testing
in the NASA Ames 11-ft subsonic and 9 × 7-ft supersonic wind
tunnels established the inlet/engine compatibility and total pressure recovery over the full YF-22 flight-test envelope. Inlet stability data measured in these tests were used to schedule the variable inlet bleed and bypass exits.[23]

8) *1/20-Scale low-speed jet-effects model:* Sting-mounted, this YF-22
model was used to gather low-speed aerodynamic/propulsion interaction data. A force balance was used to measure forces and
moments on the entire aircraft contour; oil and smoke visualiza-
tion data were also obtained. High pressure air was forced into
the model through the sting; it was then exhausted out of the
thrust vectoring nozzles. The effects of the vectored jet on low-
speed stability and control derivatives and aftbody static and dynamic pressure loads were determined over a wide AOA range.
Measured low-speed longitudinal and lateral/directional jet-
effects were used to build the database for handling qualities simulations and YF-22 flight control law development. Testing was
conducted in the Lockheed 16 × 23-ft low-speed wind tunnel.[23]

9) *1/5-Scale air data system calibration model:* Based on the results
from computational fluid dynamics methodologies, this model
was developed and used to replicate that part of the YF-22 airframe that would have an impact on the calibration of air flow
data. It consisted of the forward fuselage and included flowing
inlets. Air data calibration was accomplished by measuring local
flow angles and static and total pressures at the location planned
for the air data probe on the actual YF-22 aircraft. Forebody surface static pressures were measured at the flush air data port
sensor locations and in the vicinity of the production air data
probe. The later data were integrated with wind-tunnel testing
of an isolated air data probe. The model was tested at the
Arnold Engineering Development Center (AEDC) 16-ft supersonic and transonic wind tunnels (see Fig. 35) and in the Lockheed 16 × 23-ft low-speed wind tunnel.[23]

RADAR CROSS SECTION POLE TESTING

The Dem/Val contract required both ATF teams to conduct extensive radar cross section (RCS) testing and analyses using actual
components, computer-based RCS prediction models, and sub- and
full-scale aircraft RCS test models to ensure that their proposed EMD

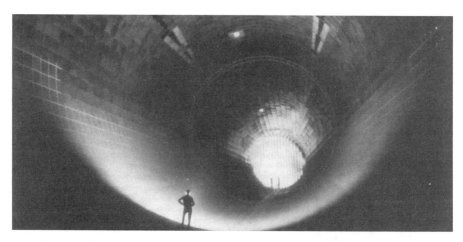

Fig. 35 Arnold Engineering Development Center Propulsion Wind-Tunnel Facility. This facility was used by both contractors for inlet, jet effects, weapons separation, and other testing.

designs would meet defined RCS requirements. The full-scale high-fidelity RCS test models were required to be equipped with all radar reflective features and absorptive materials that would be incorporated into the actual aircraft design. Their RCS was evaluated while they were mounted on a 70-ft tall low RCS pole at the Air Force Radar Target Scatter (RATSCAT) facility at White Sands, New Mexico. Prior to the RCS evaluations by the Air Force at RATSCAT, the designs had been extensively tested by both Lockheed and Northrop (in the case of the YF-22) at Lockheed's Helendale, California RCS test facility and, in the case of the YF-23, at the McDonnell Douglas Grey Butte RCS test range in California. These tests verified the RCS levels of both ATF designs. The full-scale F-23 RCS model is seen during pole testing at RATSCAT in Fig. 36.

ADVANCED TACTICAL FIGHTER DEMONSTRATION AND VALIDATION PROTOTYPE AIR VEHICLE DESCRIPTIONS

The ATF prototype air vehicles (PAVs) were relatively large aircraft that weighed in the vicinity of 60,000 lb at takeoff. Their size relative to the aircraft that they were eventually intended to replace (the McDonnell Douglas F-15 Eagle) is illustrated in Fig. 37.

LOCKHEED YF-22. The YF-22 (seen in Fig. 38) had wings with essentially full-span leading-edge flaps and very large ailerons and flaperons that occupied nearly the entire trailing edge. Leading-edge sweep angle was

Fig. 36   Full-scale F-23 radar signature measurement model.

48 deg; trailing-edge sweep was 17 deg forward. The wings used single piece thermoplastic skins over a titanium multispar structure with composite sub- and surface structures and materials (Fig. 39). The wing section was optimized for the transonic regime and had a 3.8% thickness/chord ratio. Blending at the root was for both RCS reduction and aerodynamic considerations.

The vertical tails were canted outward 27 deg and had conventional rudders; the horizontal surfaces were all-moving slab "tailerons" that functioned to provide both pitch and roll control. Their leading- and trailing-edge sweep angles matched those of the wings. All tail surfaces incorporated metal load-bearing structures and composite skins. General characteristics of the YF-22 are given in Table 2.

The YF-22 had a digital flight control system connected by fiber optic cables. A quadruple digital flight control computer had a triple-redundant signal to the hydraulic actuators that moved the control surfaces. Roll control was by a combination of differential movement of the ailerons, flaperons, and horizontal tail surfaces. Pitch control was ac-

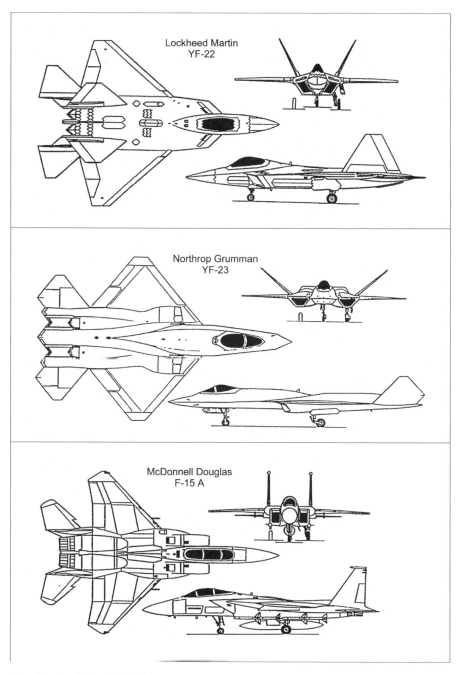

Fig. 37   Lockheed YF-22A and Northrop YF-23A PAV designs, with the standard U.S. Air Force air superiority fighter, the F-15A, illustrated for comparison.

Fig. 38   Front and side views of the first YF-22 prototype, equipped with General Electric YF120 engines, taken in conjunction with the YF-22 roll-out ceremony on Aug. 29, 1990.

complished by actuation of the horizontal tail surfaces and the use of the two-dimensional thrust vectoring nozzles. Movement of the control surfaces was executed as a function of AOA, Mach number, and landing gear position. All movable wing surfaces drooped in low-speed flight to generate desired lift.

The pneumatic air data system (PADS) consisted of two flush-mounted static ports on either side of the fuselage just forward of the cockpit and a single air data probe on either side slightly forward of them as illustrated in Fig. 40. The air data probes measured total pressure, static pressure, and angle of attack. The flush ports measured angle of sideslip—one set above the chine for low angles of attack and one set below for high AOA. At angles above 33 deg and below -5 deg,

Fig. 39   Composite section of a YF-22 wing under construction in a rotatable assembly fixture at Boeing's plant in Seattle, Washington.

Table 2   YF-22 characteristics

Wingspan	43 ft 0 in.
Aspect ratio	2.2
Overall length	64 ft 2 in.
Overall height	17 ft 8.9 in.
Total wing area	840 ft^2
Horizontal tail area	67 ft^2
Vertical tail area	109 ft^2
Empty weight	31,000 lb
Max takeoff weight	58,000 lb
Max level speed (supercruise)	1.58 Mach
Max level speed (afterburner)	2.0+ Mach
Ceiling	over 50,000 ft
g limit	+7.9

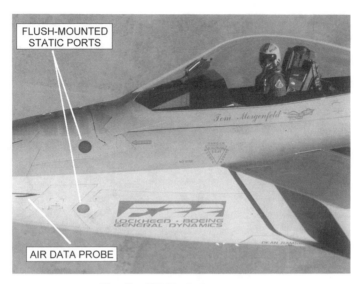

Fig. 40   YF-22 air data system.

the flight control inputs began to transition to an inertially calculated AOA. Sideslip was similarly transitioned above 60 deg and below −20 deg.

Aircraft construction was primarily metal (aluminum 33%, titanium 24%, and steel 5%) with extensive use of graphite thermoplastic composites (13%) and thermoset structures (10%).

The retractable tricycle undercarriage was stressed for no-flare landings of up to 10 ft/s. All three landing gear members were hydraulically actuated and retracted forward into the fuselage. The main landing gear bays were each fitted with an individual single-piece door, whereas the nose gear had a two-piece door. A large hydraulically actuated speed brake was located between the vertical tails; it was hinged at the forward end.

The YF-22 prototypes were powered by two Pratt and Whitney YF119-PW-100 (in the case of PAV 2) or two General Electric YF120-GE-100 (for PAV 1) advanced afterburning low-bypass ratio turbofan engines. These were in the 30,000-thrust class. Both engines were capable of self-starting and equipped with autonomous ground check-out systems. Engine hydraulic and electrical power was supplied by independent engine-mounted hydraulic and generator systems. The fixed-geometry caret-shaped air inlets were adjacent to the cockpit. The S-shaped intake ducts provided full shielding of the engine front face. Excess air was dumped through covered slots located just behind the upper intake lip; auxiliary air doors were located just ahead of these spill slots. The two-dimensional YF-22 nozzles were capable of ±20 deg pitch vectoring. Nozzle vector angle was controlled by the engine full authority digital electronic control (FADECs), which received commands from the flight control computer. Changes in nozzle exit area and thrust vectoring angles were each independently controlled with power generated by a pair of hydraulic actuators. The prototypes had fully modulated cooled afterburners. The fuel tanks were inerted using nitrogen generated by the onboard inert gas generation system (OBIGGS). An inflight air refueling receptacle was located behind the cockpit on the centerline; it rotated into position when required. The auxiliary power unit was a Sundstrand Turbomach.

An articulated ACES II zero-altitude/zero-airspeed ejection seat was fitted to both aircraft. An onboard oxygen generation system (OBOGS) provided the pilot with oxygen. The prototype YF-22 cockpit simulated an operational fighter layout and incorporated two 6 × 6-in. primary multifunction displays (MFDs) and three 4 × 6-in. secondary MFDs as seen in Fig. 41. In place of a third primary MFD, a cathode ray tube (CRT) was used to provide flight-test nose boom air data information. There were no conventional gauges in the prototype YF-22 cockpits. The active-matrix color liquid crystal displays (LCDs) were managed by finger-on-glass (FOG) touch sensitive controls. Information provided on the primary MFDs included vertical and horizontal situation display, flight-test display, equipment status display, and memory examine display. During flight testing, the one primary MFD could be replaced by a flutter excitation system panel or a stabilization recovery chute control panel. Secondary MFDs displayed the status of subsystems, fuel, stores management, and communication/navigation/

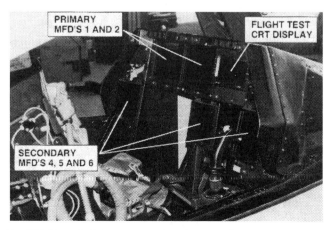

Fig. 41   YF-22 cockpit layout showing the use of multifunction displays.

identification (CDI) systems. A headup display (HUD) served as the primary flight instrument. A side-stick controller was used as part of a hands-on throttle and stick (HOTAS) arrangement; this feature enabled the pilot to accomplish all critical tasks without having to move his hands from the controls. The canopy was a single-piece laminated acrylic transparency hinged at the rear and actuated by a single hydraulic ram.

The YF-22 used common integrated processing to support all signal processing, data processing, digital input/output, and data storage functions using a single integrated hardware and software design. It used Pave Pillar-developed avionics architecture concepts and Joint Integrated Avionics Working Group-(JIAWG-) defined standards. These included the parallel-interface bus, test-maintenance bus, and data processing element, with a 32-bit central processing unit powered by an Intel 80960 reduced instruction set computer (RISC) chip. The vehicle management system (VMS) consisted of the flight control computers and bus controllers, left and right engine control computers, HUD, integrated vehicle subsystem controllers, fuel management system, pneumatic air data system transducers, inertial navigation system, mission display processors, and the integrated flight propulsion control military standard MIL-STD-1553B buses.

Data measurements were transmitted to ground stations via both encrypted and unencrypted telemetry links. Test aircraft were configured with approximately 50 accelerometers, 260 strain gauges, and 15 position sensors, as well as data from aircraft electrical signals, MIL-STD-1553B bus, HUD video, and four 16-mm 200 frame-per-s cameras for the weapon release tests. The flight-test nose boom and high accuracy digital transducers were used to measure freestream total and static pressure. This information was displayed on the cockpit CRT as indicated airspeed, altitude, rate of climb, AOA, and sideslip.

YF-23 DESCRIPTION. The YF-23 (seen in Fig. 42) had wings that incorporated one-piece leading-edge flaps and very large flaps and flaperons, occupying 90% of the trailing edge. The wings were cropped diamond-shaped with leading- and trailing-edge sweep angles of 40 deg. The all-moving ruddervators were canted outward 50 deg with leading- and trailing-edge sweeps that matched those of the wings. They were used to provide stability and control in both pitch and yaw. The YF-23 had a digital flight control system connected by fiber optic cables. Aircraft construction was primarily metal (including 35% titanium) with approximately 25% composites. General characteristics of the YF-23 are listed in Table 3.

Fig. 42   The first YF-23 seen at its roll-out ceremony on June 22, 1990.

Figure 43 shows the 10 × 15-ft cocured integral graphite skin structure that formed the upper aft-fuselage of the aircraft, forward of the nozzles.

The retractable tricycle undercarriage was stressed for no-flare landings at impact rates of up to 10 ft/s. The nose wheel retracted forward, while the main wheels retracted rearward. The YF-23 was not equipped with a conventional speed brake. To assist in reducing speed, the outboard wing trailing-edge surfaces deflected upward and the inner control surfaces downward to act as speed brakes on landing or in flight (see Fig. 44).

The YF-23 prototypes were powered by two Pratt and Whitney YF119-PW-100 (87-800) or two General Electric YF120-GE-100 (87-

Table 3   YF-23 characteristics

Wingspan	43 ft 7 in.
Aspect ratio	2.01
Overall length	67 ft 5 in.
Overall height	13 ft 11 in.
Total wing area	945 ft^2
Empty weight	37,000 lb
Max takeoff weight	64,000 lb
Max level speed (supercruise)	1.6 Mach
Max level speed (afterburner)	2.0+ Mach
Ceiling	over 50,000 ft

Fig. 43   This large graphite composite skin section covered the YF-23 engine bays. It was strengthened by longitudinal stiffeners that were cocured along with glued-on lateral stiffeners.

Fig. 44   To assist in reducing speed, the YF-23 used a combination of flap deflections to achieve high drag.

801) advanced afterburning low-bypass turbofan engines in the 30,000-lb thrust class. The fixed-geometry inverted trapezoidal air inlets were located below the wing with the wing leading edge forming the first of a two-shock system; two auxiliary air inlet doors were located above the wing on each side. Boundary layer was removed through a porous panel and vented above the wing; spill doors were located just behind the inlets. The S-shaped intake ducts curved inward and upward to provide full shielding of the engine front faces. The two-dimensional YF-23 nozzles exhausted through above-fuselage exits that were lined with heat-resistant troughs; they were not capable of vectoring. These single expansion ramp nozzles (SERNs) had a variable, external upper flap, with the lower half of the nozzle formed by a fixed ramp. The prototype troughs were made of Allison's Lamilloy diffusion-bonded titanium sheets, which were air-cooled by engine bleed air; exhaust temperatures reached 2500°F in afterburner and 1500°F in supercruise.

The YF-23 cockpit was also equipped with several multifunction displays and a large HUD. In contrast to the YF-22, it used a centrally located control column. The aircraft had a single-seat utilizing the McDonnell Douglas NACES II zero-altitude/zero-airspeed ejection system.

FLIGHT-TEST PLANNING

For the Dem/Val flight testing, the contractors were given considerable freedom to choose exactly what capabilities they would demonstrate. The only guideline was reportedly that the aircraft should be flown "for the purpose of EMD Risk Reduction."[1] It is important to note two things that the Dem/Val Prototype Air Vehicle (PAV) effort was *not:*

1) The kind of comprehensive test and evaluation or specification compliance testing that is normally done in FSD. Rather, the Dem/Val prototype flight tests were targeted to a few critical points in the flight envelope and intended to require only about 100 flight hours per design. Flight testing was to validate key ATF capabilities only (as defined by the *contractors*).

2) A competitive flyoff, but rather a data collection exercise. The contractor selected for FSD and production would be "based on the data in the FSD proposal, not on the performance of the PAVs during Dem/Val.... Their function was merely to support the projections on which the FSD proposals were based."[1]

In early 1990, the ATF SPO requested "sealed envelope" performance predictions from both airframe teams to serve as credibility criteria in the evaluation of the FSD proposals. The contractors were allowed to choose what types of data to present. The predictions were provided in June. Lockheed submitted predictions of various aerodynamic performance and flying qualities parameters, air data system functionality, engine/inlet compatibility, weapon separation trajectories, and flutter-free flight envelope.[9]

FLIGHT-TEST

From June through October 1990, all four of the Dem/Val prototype aircraft were rolled out and made their first flights. In reference to their famous twin-boom World War II fighter, the P-38 Lightning (Fig. 45), the Lockheed team dubbed their twin-tailed YF-22 the "Lightning II;" in the absence of an official name, Lightning II was often later and erroneously applied to the planned F-22 which, however, would eventually receive the official U.S. Air Force designation "Raptor" at the April 1997 EMD aircraft roll-out ceremony. Similarly, the YF-23 was often

Fig. 45   The Lockheed P-38 Lightning fighter of World War II fame.

referred to as the "Black Widow" by the Northrop team, and an hourglass shaped marking was temporarily applied to the underbelly of the aircraft. This was in recognition of the World War II Northrop P-61 Black Widow night fighter (also, like the P-38, a twin-engine, twin-boom design as seen in Fig. 46), but also was not an official designation.

The prototype aircraft were actually owned and operated by the contractors and carried civil registration (N22YF and N22YX for the YF-22s; N231YF and N232YF for the YF-23s). During a short but intensive flight-test program, all four aircraft/engine combinations achieved supersonic cruise performance on dry thrust and demonstrated many other capabilities according to their respective test plans (which were determined by the contractor teams). The Dem/Val flight-test effort was completed by the end of December 1990. Performance achieved generally matched the "sealed envelope" predictions quite closely for both teams.[9,19]

Each team's test program was conducted by a combined test force (CTF) made up of personnel from the participating airframe companies, avionics suppliers and vendors, representatives from the two engine companies, and the Air Force. The prime contractors had overall responsibility for the planning and execution of their flight-test programs, rather than the Air Force Flight Test Center (AFFTC), as is usu-

Fig. 46   The Northrop P-61 Black Widow night fighter.

ally the casc. Rather, AFFTC's primary roles were to act as facilitator
and to provide safety oversight. The YF-22A CTF composition is
shown in Table 4.[19]

The Air Force Operational Test and Evaluation Center (AFOTEC)
participated in the Dem/Val flight-test program and conducted early
operational assessments (EOAs) of the YF-22A and YF-23A's capabilities with respect to the eventual operational role of the ATF. For the
YF-22A, 11 flights totaling 13.9 h were devoted to the EOA, involving
both prototype aircraft. However, the EOA was designed to have minimum impact on the test program. "All of the AFOTEC sorties were
made up of tests that were extracted from approved test plans. However, the particular tests chosen for the EOA were more operationally
oriented," noted Richard Abrams, the Lockheed team's flight test manager.[20] Following the completion of the evaluation, AFOTEC's independent assessments were briefed up the TAC chain of command. Lt.
Col. Willie Nagle was the AFOTEC pilot for the YF-22A.

The short time available to accomplish all of the objectives that each
company had chosen dictated an aggressive and efficient flight-test approach. These objectives included ensuring that all required resources
were available to support a high sortie rate; maximizing the use of in-
flight refueling to increase the productivity of each sortie; only conducting those envelope expansion tests that were necessary to accomplish the chosen demonstration objectives; planning for and utilizing
multidiscipline test techniques; and early checkout of the AFFTC and
AFOTEC pilots who would participate in the program.[21]

Table 4   YF-22A combined test force composition

Organization	Number	Functions
Lockheed Aeronautical Systems Co.	90	Engineering, administration
Lockheed Advanced Development Co.	65	Maintenance, quality assurance, material, piloting
General Dynamics	45	Engineering, materials, piloting
Boeing Military Airplanes	40	Engineering, maintenance, quality assurance
Air Force (total AFFTC, AFOTEC)	20	Engineering, field service
General Electric	20	Engineering, field service
Pratt and Whitney	20	Engineering, field service
Total	**300**	

YF-22A FLIGHT-TEST HIGHLIGHTS

In preparation for an aggressive test schedule, manpower and resources sufficient to support a minimum of two flights per day, six days per week were located at the test site. Although exact sortie rates were expected to vary, this measure ensured that the pace of the test program would at no time be limited by lack of support. Real-time data collection and telemetry monitoring were important to maximize productivity. A secure, high-capacity network linking the Edwards Air Force Base CTF facility with Lockheed, General Dynamics, and Boeing also provided valuable capabilities including access to Lockheed's Palmdale Flight Test Data Center from the other sites and the ability to transmit flight control operational flight programs (OFPs) directly from Fort Worth to Edwards.[19]

The first YF-22 (Fig. 47), with General Electric YF120 engines, made its initial flight on September 29, 1990, piloted by Lockheed test pilot

Fig. 47   Powered by General Electric YF120 engines and flown by Lockheed test pilot Dave Ferguson, the YF-22 made its first flight on Sept. 29, 1990, taking off from Lockheed's Palmdale, California, facility and landing at the Air Force Flight Test Center at Edwards Air Force Base.

Dave Ferguson. The flight originated from Lockheed's Palmdale facility and ended at Edwards Air Force Base. The flight duration was shorter than planned because a ground station problem had delayed the takeoff, while the aircraft waited on the ramp with its engines running. Landing gear retraction was not attempted on the first flight but did not work when it was attempted on subsequent flights. In the initial setup, gear extension was hardwired, but retraction was software-controlled through the vehicle management system (VMS). The test team eventually had to hardwire the retraction function, and a successful gear retraction was accomplished on the fifth flight.[20]

Initial testing focused on airworthiness, systems functionality, and envelope expansion to Mach 1.6 at 40,000 ft (450 KEAS) to demonstrate supercruise capability. The first supersonic flight took place on the ninth flight, on October 25. The same day, on the 10th flight, Maj. Mark Shackelford became the first Air Force pilot to fly the YF-22A. Air refueling qualification was accomplished on the 11th flight on October 26 (Fig. 48). Supercruise demonstrations with the first YF-22A were conducted in early November. For the supercruise tests, the aircraft was accelerated using afterburner to the predicted supercruise Mach number, and then the engines were put into intermediate thrust (i.e., maximum nonafterburning or "dry" thrust) and the aircraft was allowed to accelerate or decelerate to its equilibrium speed. Several successful supercruise tests were conducted at altitudes ranging from 37,000 to 40,000 ft, demonstrating the ability to sustain speeds up to Mach 1.58 with the General Electric YF120 engines on dry thrust.[20]

A short downtime followed to prepare the aircraft for high angle-of-attack (AOA) testing. Preparations included installation and ground

Fig. 48   YF-22 air refueling qualification occurred on the 11th flight of YF-22 aircraft number one (Oct. 26, 1990).

checkout of a new flight control operational flight program (OFP) that would enable the use of thrust vectoring, which had not been used up to that time. The spin recovery chute was installed, and a 3-deg AOA bias in the air data system was corrected by physically drooping the air data probes together with some adjustments to the flow angle correction factors in the flight control computer. The YF-22 equipped with the spin recovery chute installation is seen in Fig. 49; the quad-mount for the spin chute canister (attached above the aft fuselage) was designed to ensure that the deployed chute would be well outside the engine exhaust plume even with the nozzles at their full-up position. The spin chute had a diameter of 28 ft; it was attached to the quad-mount with a 100-ft long riser. Several prerequisite tests were performed prior to actually starting the high AOA series. First, spin chute deployment and jettison tests were conducted during ground taxi (at 76 KIAS) and in flight (from 165 KIAS at an altitude of 25,000 ft). Flying qualities tests were accomplished up to 20-deg AOA with thrust vectoring on and off. Engine airstarts, auxiliary power unit/emergency power unit airstarts, and zero and negative g system tests were also performed. All high AOA prerequisite tests were completed by December 10, 1990.[20]

High AOA test objectives were to reach 60-deg AOA and to demonstrate good controllability by performing pitch and roll maneuvers at that condition. The test plan consisted of "a careful evaluation of aircraft stability and control/handling qualities at progressively higher angles of attack."[20] A typical test at each angle of attack consisted of the following: 1) a 1-g slowdown to the test AOA, trimming at that AOA;

Fig. 49   YF-22 number one was equipped with a 28-ft-diam spin recovery chute installation for high-angle-of-attack testing.

2) pitch, roll, and yaw doublets at the test AOA; 3) roll maneuvers; 4) throttle transients (up to 40-deg AOA only); 5) full forward stick pushovers (at intermediate thrust with vectoring on); and 6) achieving the test AOA again and repeating the pushover at idle thrust with vectoring off to determine basic aircraft pitching moment. AOA was progressively increased in 2-deg increments from 20 to 40 deg, and in 4-deg increments from 40 to 60 deg. Real-time evaluation of test results was used to issue clearances for subsequent test points. This allowed a very aggressive schedule to be maintained. The entire high-AOA test series, up to 60-deg AOA, was accomplished in 10 flights over a one-week period, from December 10 to December 17.[20]

The YF-22 was configured with multiple control surfaces to provide the control power needed to positively maneuver the aircraft anywhere within its flight envelope. As mentioned previously, the YF-22 flight control system (seen in Fig. 50) included full-span leading-edge flaps, large rudders, fully moving horizontal tails, ailerons, flaperons, and pitch-axis thrust vectoring. Roll control was provided by differential movement of the horizontal tail surfaces, flaperons, and ailerons. Although thrust vectoring was only used symmetrically, for pitch control, the use of thrust vectoring dramatically increased the roll rate capability of the YF-22 especially at lower airspeeds. This was because thrust vectoring allowed any given pitch condition to be achieved with less horizontal tail deflection, leaving more horizontal tail movement available for roll control.

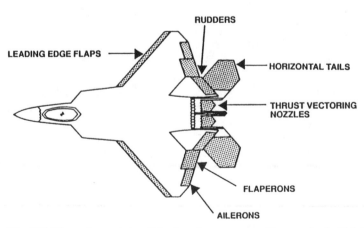

Fig. 50   The YF-22 used a variety of devices to ensure positive control at high angles of attack. These included pitch-axis thrust vectoring achieved via movable two-dimensional engine nozzles.[23]

Roll response at all angles of attack was very positive, and pilots had no difficulty holding wings level for extended periods even at extreme AOA. In the pitch axis the YF-22A was designed to be capable of pitching down from any angle of attack without the use of thrust vectoring. The idle-thrust pushovers verified this capability.[9] During the flight-test program, the YF-22 demonstrated in-flight pitch rates with thrust vectoring that were more than double those of the F-16 in the low-speed maneuvering flight regime as shown in Fig. 51.[23]

Throughout the high-AOA test program, there were no emergency deployments of the spin recovery chute. General Dynamics test pilot Jon Beesley summed up the high-AOA handling qualities of the YF-22A as follows: "It always did what I wanted it to do and never did anything that I didn't want it to do."[20]

Following the high-AOA tests, the spin recovery chute was removed. The first YF-22A subsequently accomplished further expansion of the supersonic flight envelope, performance, flying qualities, propulsion system, and loads testing out to the maximum speed of the aircraft (over Mach 2.0). The first YF-22A accomplished a total of 43 sorties and logged 52.8 flight hours.[20]

The second YF-22A, seen in Fig. 52, made its first flight on October 30, 1990, with Lockheed test pilot Tom Morgenfeld at the controls. Initial airworthiness testing was followed by preparation for the in-flight

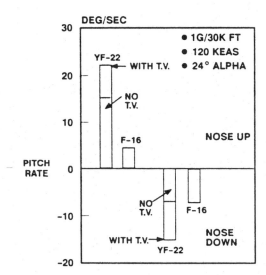

Fig. 51   YF-22 pitch rate at low speed compared to the F-16.[23]

Fig. 52   The second YF-22, powered by Pratt and Whitney YF119 engines, made its first flight on Oct. 30, 1990, flown by Lockheed test pilot Tom Morgenfeld.

missile launches. Although a "hardwired" missile launching system would have been sufficient to accomplish the planned missile launches, the second YF-22A was equipped with a software-controlled stores management system (SMS) to demonstrate and gain experience with the integration of the SMS into the total vehicle management system. As prerequisites to the missile launches, weapon bay vibration and acoustic measurements were taken. In the case of the AIM-120, this included the in-flight carriage of an instrumented missile to fully evaluate the conditions in the main weapons bay.

For both tests, the weapons bay doors were opened, and the missiles extended on trapeze launchers from the bays prior to takeoff. Figure 53 shows the EDO Corporation–developed trapeze launcher used for the AIM-120 missile test launches. An AIM-9M Sidewinder launch from the left-hand bay was performed on November 28, 1990, at the China Lake Naval Weapon Test Center in California (Fig. 54). The AIM-120 AMRAAM launch from the main bay was accomplished on December 20 at the U.S. Navy Pacific Missile Test Center near Point Mugu, California (Fig. 55). Both missiles were launched from a flight condition of Mach 0.7 at 20,000 ft. Both launches were successful, with weapon separation trajectories as predicted. Missile exhaust did not impinge on the YF-22's structure or get ingested into the YF-22's engine inlets during either the AIM-9 or the AIM-120 tests.[20]

The remainder of the second YF-22's testing included supersonic envelope expansion, supercruise demonstration with the Pratt and Whitney YF119 engine, and additional performance, flying qualities, and propulsion system testing. The highest supercruise speed sustained by

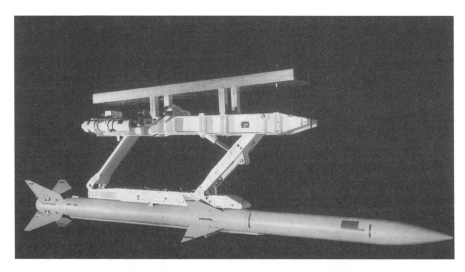

Fig. 53   The EDO-developed AIM-120 trapeze launcher.

the YF-22A with the YF119 engines was Mach 1.43. The aircraft made its 31st and final test sortie on December 28, 1990, logging 38.8 flight hours.[20]

There were no uncommanded engine in-flight shutdowns (IFSDs) or stalls on the YF-22 during Dem/Val flight testing. There was, however, a commanded IFSD of the YF120 engine, resulting in the aircraft performing a single-engine landing. This occurred on October 15, 1990, and was caused by a hydraulic leak shortly after takeoff on the third flight of the YF-22 (PAV-1). A gasket failed on the engine hydraulic system that actuated the nozzle and other engine controls. All engine hydraulic fluid was lost, causing the engine's control system logic to command a shutdown. The pilot used afterburner (for the first time in flight) on the other engine to maintain airspeed after recovering from climb-out. The pilot tried to relight the engine with both an airstart and the auxiliary power unit but was, naturally, unsuccessful. The YF-22's single-engine flying qualities were called "excellent." Investigation found the leak was in a joint on the hydraulic pump discharge line. Inspection of the joint bolt torque revealed that the four bolts that held the joint together had been torqued to a level below that required. The torque requirement was subsequently increased on all four bolt flanges.

A second hydraulic leak caused the 11th flight of YF-22 PAV-1 to be terminated on October 26, 1990. A pressure switch failed, causing loss of fluid from the number one hydraulic system. The number two system

Fig. 54   An AIM-9 Sidewinder air-to-air missile was launched from one of the side (cheek) weapons bays on the number two YF-22 on Nov. 28, 1990, over the U.S. Navy's China Lake Weapon Test Center in California.

Fig. 55   An AIM-120 AMRAAM was launched from the YF-22's main weapons on December 20 over the U.S. Navy Pacific Missile Test Center water range near Point Mugu, California.

enabled the aircraft to return to Edwards Air Force Base and land without difficulty.

The entire YF-22A flight-test effort was completed in only 91 days. Sufficient performance testing was accomplished with both aircraft to validate the performance of both the General Electric YF120 and the Pratt and Whitney YF119 types, including the performance at supersonic cruise conditions. A total of 74 test sorties were accomplished, accumulating 91.6 flight hours.[21] The YF-22A flight-test program is summarized in Fig. 56.

Relative to the "sealed envelope" predictions, the following results have been openly published[20]:

1) Supersonic cruise performance was as predicted.
2) Subsonic drag agreed with predictions and showed good agreement between the two aircraft/engine combinations (Fig. 57).
3) Supersonic drag agreed with predictions at low lift coefficients as seen in Fig. 58; there were insufficient data at high lift coefficients to make valid comparisons.
4) Sustained load factors were as predicted.
5) Specific excess power was as predicted.
6) Based on demonstrated acceleration performance, transonic drag was determined to be less than predicted.
7) Specific range was within 3% of predictions at all applicable flight conditions.
8) Maximum speed was as predicted.

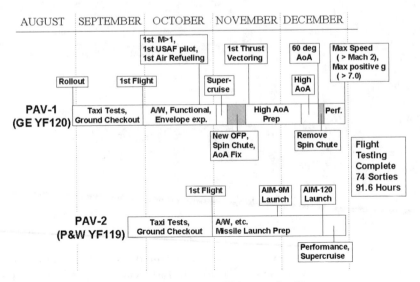

Fig. 56   YF-22A PAV flight-test timeline.

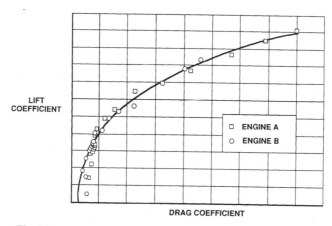

Fig. 57   YF-22 subsonic drag data comparison (Mach 0.9).

9) Maximum roll rates and time to specific bank angles were slower than predicted.
10) Maximum lift coefficient was higher than predicted.
11) Weapon bay vibration and acoustic measurements showed good agreement with predictions (which were based on wind-tunnel tests).
12) Weapon separation trajectories matched predictions, and weapon launch impact on engine operation (i.e., none) was as predicted.
13) All engine airstarts were successful; start times were equal to or less than predicted.
14) Stability derivatives agreed well with predictions from wind-tunnel testing.
15) No flutter was predicted or encountered within the YF-22 flight envelope.

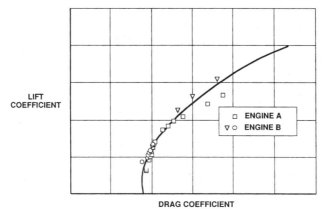

Fig. 58   YF-22 supersonic drag data comparison (Mach 1.6).

Fig. 59   The first YF-23 built was equipped with Pratt and Whitney YF119 engines. It flew for the first time on Aug. 27, 1990, piloted by Northrop test pilot Paul Metz.

YF-23A Test Highlights

The first YF-23A, seen at its roll-out ceremony in Fig. 59, was equipped with two Pratt and Whitney YF119 engines. It was transported by ground from Northrop's Hawthorne, California, facility to Edwards Air Force Base, where it was rolled out on June 22, 1990.[24] Northrop test pilot Paul Metz made the first flight on August 27.* The YF-23A accomplished its first in-flight refueling during its fourth flight on September 14 and demonstrated a supercruise capability of Mach 1.43 on September 18. The YF119-powered YF-23A also achieved a maximum speed with afterburning of Mach 1.8.[25] The aircraft was subsequently used for weapon bay tests, including acoustic measurements in the bays, carriage of an inert instrumented AIM-120, and evaluation of aircraft handling qualities with the bay doors open; although the YF-23 prototype would have been capable of testing an AMRAAM, no in-flight missile firings were conducted. The first YF-23A concluded its

---

* Metz is now with Lockheed Martin and piloted the first EMD F-22A on its first flight on Sept. 7, 1997.

flight testing on November 30 with a demonstration of its "combat surge" sortie generation capability. Six missions were flown on that day in a period of less than 10 h. The shortest turnaround time demonstrated by the YF-23 was 18 min, including simulated missile and gun rearming. Upon completion of its flight-test program, the first YF-23A had achieved 34 flights and accumulated 43 flight hours.[26]

The second YF-23A flew for the first time on October 26, piloted by Northrop's Jim Sandburg. Following initial airworthiness and functional checks, the aircraft was used primarily for supercruise, performance, and maneuverability testing. New flight control software was installed to perform automatic optimum deflection of the wing leading-edge flaps (all flights with aircraft number one were made with only two leading-edge flap positions available).[27,28] This aircraft achieved a supercruise speed of Mach 1.6 on November 29. The second YF-23A also demonstrated a 25-deg AOA capability. Although wind-tunnel tests predicted it would remain controllable up to 60 deg, the Northrop team did not choose to demonstrate this capability in full-scale flight test. Although lacking thrust vectoring nozzles, the YF-23 was able to use its huge all-moving tail surfaces for increased agility and reduced takeoff and landing distances (see Fig. 60). The second General Electric F120-powered YF-23A completed its flight testing on December 18, 1990, after 16 flights totaling 22 h.[20] Both YF-23 aircraft are seen in formation flight in Fig. 61 and during aerial refueling with a McDonnell Douglas KC-10 tanker in Fig. 62.

There was also an in-flight shutdown on the YF-23 with the General Electric YF120 engine. This incident occurred on November 6, 1990, when an instrumentation pad fitting located at the oil pump discharge tube developed a crack and began to leak high-pressure hydraulic fluid.

Fig. 60 The first YF-23 showing the large vertical tails fully deflected.

Fig. 61   Both YF-23 prototypes over the Edwards Air Force Base complex.

Fig. 62   Both Northrop YF-23s during aerial refueling operations from a McDonnell Douglas KC-10 tanker.

The hydraulic system on the YF120 flight-test engines was a self-contained system (vs a fuel actuated system planned for the F120), and the leak eventually drained the fluid reservoir. The aircraft made a single-engine landing without difficulty. Because the instrumentation pad was used for ground testing only and was not necessary for flight test, General Electric removed the pads from all flight-test engines to prevent the possibility of a reoccurrence.

The YF-23A test program totaled 65 flight hours in 50 sorties over a period of 104 days and included 7.2 h of supersonic cruise at speeds between 1.4 and 1.6 Mach. Maximum altitude achieved was 50,000 ft. Although Northrop apparently did not demonstrate this, the maximum speed of the General Electric YF120-powered YF-23 prototype would also have been over Mach 2. Highlights of the test program are illustrated in Fig. 63. Fewer details have been published about the YF-23A than about the YF-22A; however, it is known that demonstrated YF-23A performance figures generally matched the "sealed envelope" predictions and met Air Force ATF goals.

OVERALL PROTOTYPE AIR VEHICLE FLIGHT-TEST SUMMARY

The Dem/Val flight-test effort was one part of the ATF program that can definitely be described as fast-paced and aggressive. Although the total duration was only four months, both contractor teams accomplished a substantial quantity of testing and demonstrated a range of exceptional capabilities (for comparison, the YF-16/YF-17 Lightweight

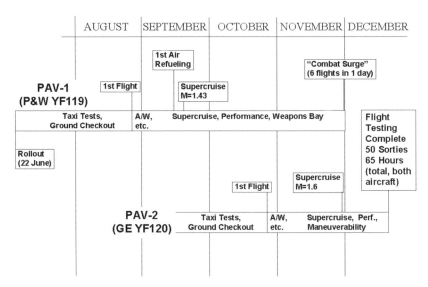

Fig. 63   YF-23A PAV flight-test timeline.

Fighter flight-test program lasted approximately one year). All major goals and predictions were met for both airframe and both engine designs. An overall summary of Dem/Val flight test events is shown in Table 5.

ADDITIONAL PROGRAM EVENTS DURING DEMONSTRATION AND VALIDATION

PROGRAM EVENTS, 1986–1989

On December 1, 1986, Col. James A. Fain became program manager. Fain had previously been the LANTIRN System program director at Aeronautical Systems Division (ASD). He managed the ATF program through the remainder of Dem/Val phase and was promoted to brigadier general in May 1989.[3]

At the beginning of Dem/Val, the contractors estimated that avionics would account for approximately 40% of the ATF's unit flyaway

Table 5   Demonstration and validation prototype air vehicle test highlights

June 1990	**"Sealed envelope" predictions provided to SPO**
June 22	YF-23A No. 1 rolled out at Edwards AFB
Aug. 27	YF-23A No. 1 first flight (YF119 engines)
Aug. 29	YF-22A No. 1 rolled out at Palmdale
Sept. 18	YF-23A No. 1 achieves supercruise speed of Mach 1.43 (YF119 engines)
Sept. 29	YF-22A No. 1 first flight (YF120 engines)
Oct. 26	YF-23A No. 2 first flight (YF120 engines)
Oct. 30	YF-22A No. 2 first flight (YF119 engines)
November	**EMD RFP issued**
Nov. 3	YF-22A No. 1 achieves supercruise; highest supercruise speed with this aircraft Mach 1.58 (YF120 engines)
Nov. 15	First thrust vectoring on YF-22A No. 1
Nov. 28	AIM-9M missile launch from YF-22A No. 2
Nov. 29	YF-23A No. 2 achieves supercruise speed of Mach 1.6 (YF120 engines)
Nov. 30	34th and final flight of YF-23A No. 1 (43 h)
Dec. 10	YF-22A No. 1 begins high AOA testing
Dec. 17	YF-22A No. 1 completes high AOA testing
Dec. 18	16th and final flight of YF-23A No. 2 (22 h); YF-23A flight testing completed
Dec. 20	AIM-120 missile launch from YF-22A No. 2
Dec. 27	YF-22A No. 2 achieves supercruise speed of Mach 1.43 (YF119 engines)
Dec. 28	YF-22A No. 1 achieves max Mach (over 2) and max positive g (over 7) (YF120 engines); YF-22A flight testing completed
Dec. 31	**EMD proposals delivered by both contractor teams**

cost. They also identified avionics as one of the greatest risk areas. The ATF SPO was directed to develop an avionics management plan as one of the conditions for final Milestone I approval. This plan provided overall guidance to the ATF avionics efforts during Dem/Val. As noted, the Dem/Val plan had been structured all along with a high level of emphasis on avionics, and the avionics prototypes (ground and flying) constituted one of the three major thrusts of the ATF Dem/Val phase.

Early in Dem/Val a new cost and operational effectiveness analysis (COEA) was undertaken. A COEA had been performed in 1986, but it was fairly specific in its scope and did not address several important issues. The 1987 COEA was more comprehensive and correlated the results of all major ATF studies from the 1981 RFI onward.[15]

In February 1987, the ATF SPO assumed management of the ATF engine effort. Personnel formerly managing the engine program in the New Engine SPO (ASD/YZ) moved into the ATF SPO. During 1987, several Air Force Wright Aeronautical Laboratories (AFWAL) avionics programs and their funding were transferred into the ATF SPO. In August 1987, the ATF SPO itself was transferred out of the Deputy for Tactical Systems (ASD/TA) and became its own two-letter office, ASD/YF (no longer under an ASD Deputy). By the end of 1987, there were 168 personnel assigned to the SPO. In August 1988, the ATF SPO also absorbed the integrated electronic warfare system (INEWS) program, which had previously been managed by the Electronic Combat and Reconnaissance SPO (ASD/RW).[3]

In early 1988, the SPO came under some pressure to increase the extent of competition in the program (not for the first time). This was part of a DOD-wide effort to maximize competition in defense contracting and to dual-source as many items as possible. With some smaller systems, one possible acquisition strategy is that contractors who team during development are assigned "leader/follower" roles in production. Both contractors develop the capability to produce the system in its entirety. Shares of the initial production lot(s) may be assigned at the beginning of production, but the contractors sometimes compete against each other for shares of subsequent lots. With very large and complex systems such as entire aircraft, this strategy is not practical, and it was never the plan for ATF. As early as 1984, it had been reported that

> The Air Force insists it will choose only one of these seven [contractors participating in the concept development contracts at that time] as prime contractor.... "There may be associate contractors on the ATF," predicts Karl G. Harr Jr., president of the Aerospace Industries Assn. Two or more, he adds, might wind up helping the

victor produce the plane [by taking responsibility for various tasks, components, or sub-assemblies].[29]

However, there was not any plan to dual-source the production of the airframe itself or major components thereof. In 1988, the Air Force announced that although there would only be one source for the airframe, there would be dual sources for many subcontract components.[1]

An important activity during Dem/Val was planning for the EMD and subsequent phases. The first overall ATF acquisition program baseline (APB) was prepared during 1988. The APB lays out the acquisition strategy for a system and establishes cost, schedule, and performance thresholds for the program. It is effectively a contract between the program manager and the applicable (service or DOD) acquisition authority. The APB was approved by Lt. Gen. Thurman (ASD commander) and forwarded to the Air Force Service Acquisition Executive, John J. Welch, for signature in May 1988. It was then sent to the Defense Acquisition executive [i.e., the undersecretary of Defense (Acquisition)] who signed it in December 1988. The APB and many other supporting acquisition plans and documents evolved during Dem/Val in response to evolving system requirements, changing DOD acquisition policies and procedures, funding pressures, etc. During 1989, the Air Force changed the overall acquisition strategy for full-scale development (FSD) and production from fixed price plus incentive fee (FPIF) to cost plus award fee (CPAF).[3]

In early 1989, the Air Force announced that the first FSD flight would be delayed by one year, from 1992 to 1993, to 1) allow more time to take advantage of maturing technology, 2) reduce concurrency, and 3) trim near-term budgets. There were "numerous unconfirmed reports around that time" that both teams were having trouble meeting weight and cost goals [50,000-lb maximum takeoff weight (MTOW) and $35 million FY85, respectively, at that time].[1] According to one Lockheed source, "We have a sliding scale ... we move the index up and down the scale to show what has to be given up to keep the weight at 50,000 lb. We're giving the customer a lot of choices, a lot of decisions to make."[30]

In mid-1989 the House Appropriations Committee cut ATF funding from the FY90 Appropriations Bill, reportedly as a "warning to the Air Force to clarify Congressional concerns over the ATF's acquisition strategy, costs, and technical risks. The Committee criticized the program for overly optimistic cost and production rate estimates, schedule slips, software problems, and plans to sign production contracts before FSD is finished."[1] Full funding was eventually restored, but the incident kept all these issues at a very high level of visibility. Another 1989 report from the House of Representatives accused DOD of not being serious about the eventual cross-service use of the ATF and ATA.[1]

On August 18, 1989, the draft RFP for full-scale development [FSD; soon thereafter renamed engineering and manufacturing development (EMD)] was released to the weapon system and engine contractors for comment.[3]

Also in mid-1989, Lockheed announced that completion of the YF-22A would be delayed by an earlier decision to perform a substantial redesign for weight reduction (the mid-1987 to early 1988 redesign[9]). The Air Force agreed to a six-month postponement of flight testing, and Northrop stretched its program the same amount so as to avoid a break in its ATF effort. In December 1989, Northrop reduced its company funding commitment to the program. As noted earlier, the actual first flights for the YF-23A and YF-22A were August and September 1990, respectively.[1]

MAJOR AIRCRAFT REVIEW AND ALTERNATIVES TO THE ADVANCED TACTICAL FIGHTER

By this time the events in the Soviet Union and elsewhere in Eastern Europe were indicating a dramatic change in the primary threats to U.S. security. Accordingly, Secretary of Defense Richard Cheney launched the major aircraft review (MAR) in January 1990 to revisit aircraft procurement decisions involving the ATA, the B-2, the C-17, and the ATF and to assess tradeoffs between new systems and upgrades to existing systems against a scenario of two nearly simultaneous major regional conflicts (MRCs). Results of the MAR were announced in April. Secretary Cheney stated,

> We can afford to slow down the pace of developing and fielding the next generation of aircraft, in some cases can reduce the planned buy, and can extend existing systems longer than originally planned. We cannot cancel or halt modernization programs, but we can slow some programs and save some money....[31]

Consistent with these statements, the ATF requirements were not changed as a result of the MAR. The planned EMD start date of mid-1991 was maintained, but a low rate initial production (LRIP) decision was deferred four years until 1996[32] with completion of the production run in 2014. The peak production rate was reduced from 72 to 48 per year, with this rate to be reached in 2001.[1] The estimated unit flyaway cost (UFC) increased from $41.2 to $51.2 million (in FY90 dollars) due to these changes, plus the use of more detailed estimates and a better understanding of actual F-22 requirements (compared to early Dem/Val).[32]

At approximately the same time as the MAR, Defense Intelligence Agency (DIA) officials noted that

> even though the number of Soviet forces are decreasing, the capabilities of the remaining Soviet forces will continue to be a formidable threat. New fighter aircraft continue to enter the inventory, and other military equipment continue to be modernized. In addition, the spread of high-technology weapons to many other countries presents a new and more sophisticated global threat to U.S. forces.[14]

Nevertheless, the MAR was criticized for failing to address issues of affordability; for relying on data provided by the services to assess their aircraft programs; and for failing to consider such alternatives as the "Falcon 21," an F-16 derivative then being proposed by General Dynamics. F-15 and F-16 derivatives were proposed repeatedly (before the MAR and since) as cheaper alternatives to the ATF.

> One senior Pentagon official has noted that the Falcon 21 could provide real competition to the ATF, since the ATF stores all its weapons internally, limiting its payload. The semi-submerged [i.e., conformal] weapons carriage configuration of the Falcon 21 would give it more flexibility in weapons carriage, while still providing a degree of low observability.[33]

However, "a degree of low observability" is not sufficient to achieve acceptable survivability in a high threat environment. Radar detection range varies only as the fourth root of the target's radar cross section (RCS). Thus, for example, to reduce the range at which a fighter can be detected by a factor of 10, its RCS must be reduced by a factor of 10,000. For this reason, necessary reductions in detection range simply cannot be achieved through modification or retrofit of existing aircraft designs. Derivative aircraft, while appealing economically, simply cannot provide the survivability, and resulting ability to achieve air superiority at the time and place of the U.S. theater commander's choosing, that the ATF will provide.

The bottom line is that the ATF survived the MAR, with a stretched program but no change to the basic weapon system requirements. However, just three months after the MAR was completed, the Air Force was directed to evaluate the Falcon 21 and a proposed "F-15XX" as alternatives to the ATF. Findings presented to Congress in June 1991 again indicated that these alternatives could not fill the ATF's role. The bottom line was that these aircraft would have had one third the capability at two thirds the development cost of the F-22. Nevertheless, pressure to consider alternatives to the ATF has continued and will probably continue until the last F-22 comes off the production line.

As of early 1990, the projected costs of the ATF program in then-year dollars were Dem/Val—$3.8 billion; EMD—$10.53 billion; production—$65.08 billion; total—$79.4 billion. This was a $12.3 billion increase since May 1989. The increase was attributed to higher inflation rates than predicted previously, combined with slips in the program (the six-month Dem/Val slip that was announced in 1989, and the changes that resulted from the MAR).[14]

TRANSITION TO ENGINEERING AND MANUFACTURING DEVELOPMENT

As noted earlier, the first draft RFP for FSD [shortly thereafter renamed engineering and manufacturing development (EMD)] was released on August 18, 1989. The second draft RFP for the EMD phase was released in April 1990. The final version was issued on November 1, 1990. The PAV flight-test programs were completed in December 1990, and the airframe contractors submitted their EMD proposals on December 31 (only three days after completion of the Lockheed team's Dem/Val flight testing). Lockheed's proposal was reportedly 20,000 pages long and was delivered aboard a specially chartered Convair transport aircraft.[19] The engine EMD proposals were turned in on January 2, 1991. The airframe teams made oral summary presentations to the program office on the following day.[3] While in source selection, consideration of the Naval Advanced Tactical Fighter (NATF) was officially dropped.[1]

On April 23, after slightly more than three months to review the Dem/Val results and EMD proposals, Secretary of the Air Force Donald B. Rice announced that the Lockheed team had been selected for the F-22 EMD phase and that Pratt and Whitney was selected as the winner of the ATFE competition. The contract win was attributed to their superior technical proposals and program management plans. An Air Force assessment also concluded that the proposed Lockheed and Pratt and Whitney concepts were lower risk and slightly lower cost. Rice said the two winning designs "clearly offered better capability at lower cost, thereby providing the Air Force with a true best value."[34]

According to one source selection official,

> Lockheed and Pratt and Whitney were rated higher in their technical proposals and their plans for managing the development program...the main factor in each of these ratings were the Air Force's assessment of risk. Lockheed and Pratt were considered more likely to accomplish what they proposed, and to manage the

development program successfully. Northrop and GE were considered more likely to have problems...the Air Force also concluded that the Lockheed–Pratt and Whitney aircraft would cost less than the other three combinations of airframe and engine. The difference was not great...but a few percentage points in as high-cost a program as the advanced tactical fighter (ATF) involve large amounts of money.[34]

The Northrop F-23 design was reported to be slightly faster and "stealthier," whereas the F-22 was more optimized for maneuvering. None of these differences, however, was decisive, as both designs met the stated requirements for the ATF. The F-22 was more like a conventional fighter design and therefore appeared to represent less uncertainty. Perhaps more important than any design differences were the management and risk mitigation plans contained in the proposals. Lockheed's proven experience with the F-117A and Boeing's avionics and aerospace manufacturing expertise may have been factors in that team's favor.[1]

As with the aircraft, the winning Pratt and Whitney F119 engine was more conventional than the General Electric F120. The F120 was the first variable-cycle turbofan fighter engine ever to fly. The F120 might have had more ultimate performance potential, but it also might have cost more to develop and had a higher perceived level of technical risk.[1,5]

The ATF program had started out with a great deal of willingness to utilize advanced technologies and concepts to achieve a weapon system that would ensure air dominance for the foreseeable future. However, by 1991, cost and risk had become important enough to influence the source selection decision, given that both competing engine designs and both competing weapon system concepts had demonstrated the capability to meet Air Force performance objectives.

In late July 1991, the Defense Acquisition Board (DAB) announced the Milestone II decision (EMD Start) on the ATF program. Some changes were made at this time. The planned production quantity was officially reduced from 750 to 648, scheduled to be completed in 2012 (two years earlier than the previous plan). System requirements were changed to increase the level of avionics integration and to add a two-seat trainer variant of the F-22 (the F-22B). These changes were estimated to add $4.0 million and $1.2 million, respectively, to the F-22's unit flyaway cost, bringing the estimated UFC to $56.9 million.[32]

The EMD contracts were awarded on August 3, 1991. The total initial value was approximately $11 billion: $9.55 billion to Lockheed and $1.375 billion to Pratt and Whitney. The contracts were cost plus award fee (CPAF) with a 4% initial fee plus up to 9% in performance-based fees. Acquisition regulations by this time did not permit major development contracts to be fixed-price, as the Dem/Val contracts had been. The overall reimbursable costs were capped at $12.7 billion.[1]

According to the original plans for the EMD phase, 33 F119-PW-100 engines and 13 F-22 airframes were to be delivered—nine single-seat and two two-seat flightworthy aircraft, plus one static test and one fatigue test airframe. The F-22 outer mold line was frozen in October 1991. Fabrication of the first EMD aircraft was to begin in December 1992. Peak level of EMD effort would occur in 1994, and an LRIP decision would be made in 1996. In addition to the 11 flightworthy EMD aircraft and two static/fatigue test aircraft, four "preproduction verification" F-22As were to be used for initial operational test and evaluation (IOT&E). Full production was planned to begin in late 1997 or early 1998. The F-22A design that emerged is compared with the YF-22 in Fig. 64.

In January 1991, the F-22 team headquarters was relocated from Burbank, California, to Marietta, Georgia. This move resulted in significant personnel attrition to the F-22 team—one Air Force program office source reported that, although the company hoped to retain 70%

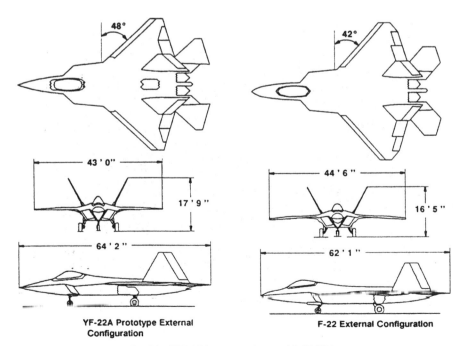

YF-22A Prototype External Configuration

F-22 External Configuration

Fig. 64   YF-22A comparison with F-22A.

of the personnel following the move, the number that actually moved with the program from California to Georgia was closer to 30%. As noted, the production run was expected to consist of 648 aircraft at a peak production rate of 48 per year. By 1992, a 190,000-ft^2 automated production facility for the F-22, was under construction within the main production building at Lockheed's plant in Marietta, Georgia. A 138,000-ft^2 addition to Lockheed's Adhesive Bonding plant in Charleston, South Carolina, was also under construction.[19] These facilities and many others at various locations had to be built well ahead of the ramp-up to full rate production and were sized to the expected production rate as of the time the facilities were designed. Thus many of the costs related to production were locked in fairly early. Nonrecurring start-up costs are included in the UFC of the F-22; this gives some indication why every cut in the production rate or quantity has an adverse impact on the F-22's UFC.

On October 30, 1991, the second (Pratt and Whitney powered) YF-22A began flight tests in support of the EMD effort. One hundred hours of testing were planned to be accomplished in approximately 25 sorties. However, a landing accident occurred on April 25, 1992, after over 60 h of EMD-related testing. Following a series of pitch oscillations starting at roughly 40 ft above the runway during landing approach, the aircraft hit the runway with the landing gear up, slid approximately 8000 ft, and caught fire. The pilot left the aircraft unhurt, but the aircraft was severely damaged; further flying of the YF-22 was discontinued. Most of the objectives of the post-Dem/Val test effort had been accomplished. The aircraft was rebuilt, but not to flying status. Instead, the aircraft was reconfigured to be representative of the EMD configuration and was later used for antenna testing at the Rome Air Development Center, Griffiss Air Force Base, New York (as seen in Fig. 65).[21]

The landing incident led to an examination of the control laws and the control law design methodology used on the YF-22. The large number of different control effectors, including thrust vectoring, provide tremendous control authority and responsiveness, but the proper management of such authority requires careful attention.

Lockheed had also proposed using the Boeing 757 Avionics Flying Laboratory (AFL) for continued testing early in the EMD phase. However, the avionics had been removed from the aircraft, and it was determined then that the desired data could be obtained more cost-effectively through ground tests, analysis, and simulation, rather than reinstalling the already outdated Dem/Val avionics in the 757 testbed aircraft. However, the Boeing 757 AFL is again being used as a flying avionics integration laboratory in support of the F-22 program. It has

Fig. 65   The reconfigured YF-22 used in support of EMD avionics testing.

now been extensively modified to support F-22 EMD program objectives and is now equipped with more mature F-22 EMD avionics systems rather than the early Dem/Val prototype avionics.

DEMONSTRATION AND VALIDATION SUMMARY

The following statistics give some idea of the scope of effort that was undertaken in the formal ATF Dem/Val phase in support of the three major Dem/Val tasks. These figures represent only the Lockheed-led team; a generally similar level can be assumed for the Northrop/McDonnell Douglas team effort[9,35]:

1) *Wind-tunnel tests:* 18,000 h involving at least nine different models (numerous configuration variations tested with each model).

2) *RCS tests:* 3200 h.
3) *Design/analysis:* 10,000,000 h.
4) *Avionics flying laboratory tests:* 159 h.
5) *Prototype demonstrations:* five major avionics ground demonstrations.
6) *Manned simulations:* 1100 h.
7) *RM&S demonstrations:* 400 distinct demonstrations.
8) *P&W F119 engine ground testing* [supportive of both teams' PAV efforts]: 110,000 h of component tests and 3000 h of full-up engine tests.
9) *YF-22A flight test:* 91.6 h.

The Lockheed/Boeing/General Dynamics team invested $675 million in company funds during Dem/Val, whereas the Northrop/McDonnell Douglas team invested approximately $650 million. Each engine company also invested approximately $100 million. The total costs of the Dem/Val program (not including government costs, such as AEDC testing) are summarized in Table 6.

Note that the amounts in Table 6 reflect final contract values in fiscal year 1990 dollars, which account for the increases relative to the original contract values at the time awarded. The Dem/Val prime contracts (airframe + critical subsystems), for example, were originally $691 million each at time of contract award, while the final values as shown in Table 5 were approximately $817 million each in fiscal year 1990 dollars for a total of $1633 million. The engine contracts include the original $202 million JAFE contracts, plus $30 million interim/long-lead added in 1986, plus $341 million full prototype restructuring added in

Table 6 Demonstration and validation costs[36]

Description	Cost (FY90 $M)
Government funding	
Airframe	1097
Critical subsystems (avionics)	536
JAFE/ATFE	1946
Simulator	4
INEWS/ICNIA	241
Government subtotal	3824
Company contributions	
Airframe	1325
Engine	200
Company subtotal	1525
Total Dem/Val	**5349**

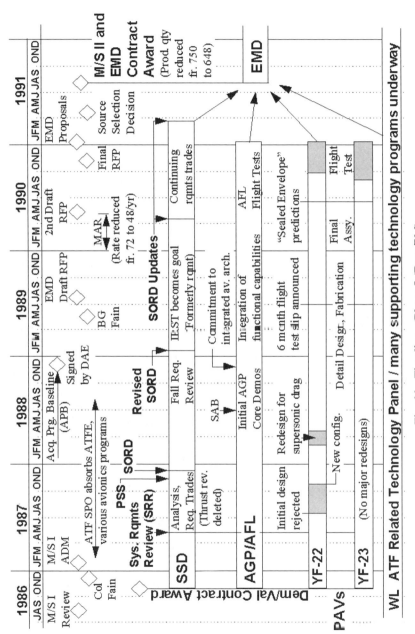

Fig. 66   Timeline—Phase I (Dem/Val).

1987, plus $290 million EMD long-lead protection contracts awarded in 1990, for each of the two engine companies, plus inflation and some minor adjustments/additions.

A timeline for the Dem/Val phase of the ATF program is presented in Fig. 66.

REFERENCES

[1]Twigg, J. L., *To Fly and Fight: Norms, Institutions, and Fighter Aircraft Procurement in the United States, Russia, and Japan,* Ph.D. Dissertation, Dept. of Political Science, MIT, Cambridge, MA, Sept. 1994.

[2]Dornheim, M. A., "Lockheed Team Will Test ATF Cockpit in Boeing 757 Flying Laboratory," *Aviation Week and Space Technology,* Nov. 10, 1986.

[3]Ferguson, P. C., *Advanced Tactical Fighter/F-22 Annotated Chronology* (Draft), ASC/HO, Wright–Patterson AFB, OH, Aug. 1996.

[4]Various Weekly Activity Reports (WARs), from ATF System Program Office to the Commander, Aeronautical Systems Div., late 1987.

[5]Sweetman, B., *YF-22 and YF-23 Advanced Tactical Fighters: Stealth, Speed and Agility for Air Superiority,* Motorbooks International, Osceola, WI, 1991.

[6]Rawles, J. W., "ATF Program Gathers Momentum," *Defense Electronics,* July 1988, pp. 43–51.

[7]Moorhouse, D. J., *Lessons Learned from the STOL and Maneuver Technology Demonstrator,* Rept. No. WL-TR-92-3027, Flight Dynamics Directorate, Wright Laboratory, Wright–Patterson AFB, OH, June 1993.

[8]Operational Requirements Document (ORD), *F-22 Advanced Tactical Fighter.* SECRET. (Unclassified information only used from this source.)

[9]Mullin, S. N., "The Evolution of the F-22 Advanced Tactical Fighter," Wright Brothers Lecture, AIAA Paper 92-4188, Washington, DC, Aug. 1992.

[10]ORD Attachment: *Requirements Correlation Matrix.* SECRET. (Unclassified information only used from this source.)

[11]WAR, March 10–16, 1988.

[12]Schultz, J. B., "New Air Force Fighter to Employ Stealth Design and Advanced Electronics," *Defense Electronics,* April 1984, pp. 97–102.

[13]"Fighter Weapons," *Aviation Week and Space Technology,* Letter to the Editor, Jan. 28, 1985, p. 108.

[14]*Defense Acquisition Programs: Status of Selected Programs,* General Accounting Office, GAO/NSIAD-90-159, June 1990.

[15]Borky M. J., Col. USAF, *Press Release on Avionics for the Advanced Tactical Fighter (ATF),* ASD 90-2347, Aug. 23, 1990.

[16]Borky, M. J., Col. USAF, *Advanced Tactical Fighter Demonstration/Validation Phase Avionics Program,* Press Conference Briefing, Aug. 23, 1990.

[17]"YF-23 Advanced Tactical Fighter Prototype: A Revolution in Air Superiority," Northrop Corp., Hawthorne, CA, Feb. 1991.

[18]WAR, Oct. 20–26, 1988.

[19]Abrams, R., and Miller, J., *Lockheed (General Dynamics/Boeing) F-22*, Aerofax Extra No. 5, 1992.

[20]Abrams, R. (Director of Flight Test, Lockheed Advanced Development Company), *YF-22A Prototype Advanced Tactical Fighter Demonstration/Validation Flight Test Program Overview*, 1991.

[21]Miller, J., *Lockheed's Skunk Works: The First Fifty Years*, Aerofax, Arlington, TX, 1993.

[22]Piccirillo, A. C., Col. USAF (Ret.), Personal notes, 1983–1986.

[23]Glasgow, E., "YF-22 Supermaneuverability," Draft Paper, Lockheed Corp., Burbank, CA, Undated.

[24]Scott, W. B., "YF-23 Previews Design Feature of Future Fighters," *Aviation Week and Space Technology*, July 2, 1990, pp. 16–21.

[25]"YF-23A Conducts Air Refueling Tests, Reaches Supersonic Speeds on Following Mission," *Aviation Week and Space Technology*, Sept. 24, 1990, p. 25.

[26]"First YF-23 Completes Dem/Val, Demonstrates 'Surge' Capability," *Aviation Week and Space Technology*, Dec. 10, 1990, p. 83.

[27]Scott, W. B., "YF-23 Prototype Undergoes Flutter Tests, May Finish Flight Evaluations Next Month," *Aviation Week and Space Technology*, Oct. 22, 1990, pp. 114–115.

[28]Scott, W. B., "ATF Contractor Teams Accelerate Evaluations with Four Prototypes," *Aviation Week and Space Technology*, Nov. 5, 1990, p. 98.

[29]"Breaking Barriers to Build the Next Air Force Fighter," *Business Week*, Feb. 13, 1984, p. 128.

[30]"Contractors Struggling," *Armed Forces Journal International*, Feb. 1989.

[31]*Department of Defense Major Aircraft Review, Hearings on National Defense Authorization Act for FY91*, before the House Armed Services Committee, 101st Congress, 2nd Session, April 26, 1990, pp. 692, 693.

[32]Walley, R., Maj. USAF, *Information Paper: History of F-22 Unit Cost Estimates*, Office of the Assistant Secretary of the Air Force (Acquisition), Fighter Div. (SAF/AQPF), March 21, 1996.

[33]Brown, D. A., "General Dynamics Evaluates Concepts for F-16 Successor," *Aviation Week and Space Technology*, June 11, 1990, pp. 21–22.

[34]Bond, D. F., "Risk, Cost Sway Airframe, Engine Choices for ATF," *Aviation Week and Space Technology*, April 29, 1991.

[35]*Read-Ahead Blue Book* for the Defense Acquisition Board Advanced Tactical Fighter Milestone II Review, June 25, 1991.

[36]"Annex B—Program Life-Cycle Cost Estimate Summary ($M)," (Congressional Budget C-Document), Office of the Assistant Secretary of the Air Force (Acquisition), Fighter Div. (SAF/AQPF), Aug. 1, 1991.

Chapter 5

RELATED TECHNOLOGY AND SUBSYSTEM DEVELOPMENTS

INTRODUCTION

The Advanced Tactical Fighter (ATF) program has always been closely associated with the development of key aeronautical technologies. The Milestone Zero decision in 1981 specifically endorsed a strategy of focused technology thrusts prior to the construction and flight test of prototype aircraft. In fact, the program name for the early part of the concept exploration phase, as set forth in the August 1982 program management directive (PMD), was Advanced Tactical Fighter Technologies.

In studies of the early 1970s, technology assessments were typically an important part of each effort. The air-to-surface studies of 1976–1978 identified important technology areas and developed technology road maps in each area.

Although the ATF mission was envisioned as air to-ground (A-G) during that period, many of the top-level weapon system attributes would be equally necessary for an air-to-air (A-A) fighter operating in the threat environment of the 1990s and beyond. These included long range; integrated avionics, supersonic persistence, advanced countermeasures, and reduced observables for improved survivability and lethality; and improved reliability, maintainability, and supportability (RM&S) to achieve high sortie rates from austere and/or damaged bases. Many of the supporting technologies therefore remained applicable as the mission shifted from A-G to A-A during 1980–1982.

The major change in technology thrusts, as the primary mission shifted from A-G to A-A, was in the area of offensive mission systems. The major mission systems (avionics and weapons integration) technology needs for air-to-ground were improved targeting and weapon delivery to allow higher speed (i.e., supersonic) attack profiles, multiple kills per pass, and night/adverse weather capability. As the ATF's mission shifted to air superiority, the need shifted toward those technologies needed for beyond-visual-range air combat, such as improved identification friend or foe (IFF). In either case, avionics integration would be a significant part of the necessary avionics technology development.

169

In 1981, Wright Laboratories (WL) established the ATF Technologies Committee (later renamed the ATF Related Technologies Panel), which has served continuously since then to coordinate WL technology programs with ATF technology needs. From 1983 through 1986, WL assigned a deputy for Technologies to the ATF SPO (Bill Goesch).[1]

In 1983, based on an analysis of the ATF request for information (RFI) responses, the ATF Technologies Committee identified 23 technology programs that were critical to the development of the ATF. This list, which follows and is not in priority order, was used to restructure Wright Laboratories' technology priorities to support the ATF's technology needs.[1]

1) *Avionics*
   Integrated Avionics Arch. (Pave Pillar)
   ICNIA
   INEWS
   Ultra Reliable Radar (URR)
   VHSIC Common Signal Processor
   VHSIC 1750 Computer
   IR Search and Track System (IRSTS)
   Air-to-Air Attack Management
   Non-cooperative Identification
   Integrated Control Avionics for Air Superiority
2) *Cockpit/Crew Systems*
   Mission Integrated Transparency System
   Crew Escape System Technology
   Cockpit Automation Technology
3) *Flight/Propulsion Control*
   Self-Repairing Flight Control
   STOL and Maneuver Technology
4) *Propulsion*
   Advanced Turbine Engine Gas Generator
   Aircraft Propulsion Subsystem Integration
5) *Structures and Materials*
   Graphite Epoxy/Thermoplastics
   Advanced Metallic Structures
   Carbon-Carbon Structures
6) *Subsystems/Miscellaneous*
   Survivability/Laser Hardening
   Non-Flammable 8000 psi Hydraulics
   Atmospheric Electrical Hazard Protection

In 1985, and periodically thereafter, the list of WL technology programs was updated and revised to support the technology needs of the evolving ATF. Formal technology transition plans were developed, with the ATF program as the customer. WL programs were complemented by industry internal research and development (IR&D), as well as subcontracts from the ATF primes to the developers and suppliers of materials, equipment, and subsystems. Through the combined efforts of the ATF SPO, Wright Laboratories, and industry, a large number of technologies have matured and have been transitioned into the ATF, as well as into other weapons systems. Some of the most significant ATF-related technology developments will be discussed next. Because of the large investment in the ATF engine development (approximately half of the entire government funded ATF effort), the evolution of the propulsion system is described in detail separately, in Chapter 6. Supporting propulsion technologies (such as ATEGG and APSI), however, are described in the next section.

AVIONICS

One of the essential features of the ATF/F-22 is integrated avionics. At many levels, the ATF avionics are more integrated than any other fighter aircraft. In the past, the introduction of progressively more advanced and more numerous subsystems for sensing, communications, fire control, countermeasures, etc., on fighter aircraft has led to a number of problems. First, the pilot has been left with the task of receiving and prioritizing the data coming to him from the various systems and of providing the necessary control inputs to each one. Second, each new function is often performed by a new piece of dedicated hardware. This must be put into a new "hole" somewhere on the aircraft, with power, cooling, and an interface to the central computer. It also adds a new set of spare parts and maintenance procedures. As a result, avionics have become a major driver of maintenance and support costs and logistic footprint. The ATF program set out to reverse this so-called federated approach, with its inherent difficulties, by planning, developing, and designing the ATF's avionics in an integrated manner.

The technology of integrated avionics includes fault tolerant architecture, common avionics hardware modules to perform a variety of functions (minimizing the amount of task-specific hardware), shared resources, and greatly improved cockpit automation relative to earlier aircraft. In the F-22 each display presents information from a variety of sensors (this is referred to as sensor fusion). The amount of unique sensor processing equipment is reduced by common processors that receive data from various physical sensors. The sensors themselves—the antennas, apertures, etc.—are integrated and combined wherever possible. The overall architecture is integrated, and a small number of common processing modules form the "building blocks" for the different avionics systems.

This approach greatly reduces the number of unique hardware spare components that must be stocked to support the ATF. It also allows functional redundancy to be achieved without the physical duplication of every piece of equipment on the aircraft. Instead, one common module serves as a backup for several others, protecting against failure of any one. After the backup module(s) becomes fully committed, additional failures are accommodated by reconfiguring the system to utilize the remaining, functional modules for the highest priority tasks. This concept underlies many areas of ATF avionics, including core processing as well as the individual functional areas such as electronic warfare (EW), communication, navigation, and identification (CNI), etc.

Some of the other benefits of integrated avionics include weight/volume savings, increased availability, enhanced crew/weapon effectiveness, and reduced cost of ownership. Increased commonality, among components on the ATF and between the ATF and other systems, could ultimately also lead to economies of scale in production and reduced acquisition costs. (Avionics are expected to account for roughly 40% of the unit flyaway cost of the F-22.[1,2])

The overall integrated avionics architecture was developed in a program called Pave Pillar. Other programs focused on the integration of specific sets of functions and/or other types of improvements to the avionics. A few of the most important ones are discussed in the following sections: Integrated Electronic Warfare System (INEWS), Integrated Communication, Navigation and Identification Avionics (ICNIA), and Ultra Reliable Radar (URR).

PAVE PILLAR

Pave Pillar started in the early 1980s as an Air Force Avionics Laboratory demonstration program to develop and demonstrate integrated avionics architecture for future combat aircraft. Pave Pillar focused on the use of common avionics modules to perform a variety of functions, rather than using a large number of task-specific hardware components. The modules were to be based on very high-speed integrated circuit (VHSIC) technology to achieve high performance, high reliability, and compact packaging.

VHSIC, in turn, was a DOD-sponsored program intended to develop high-speed, common, readily adaptable avionics equipment for all of the military services. VHSIC was motivated to a large extent by various advances in Soviet acquisition and tracking capabilities that, as reported in 1984,

> ...suggest an increasing Soviet mastery of digital electronics, a technology in which the U.S. has begun to lose its longtime lead.
>
> The urgent need to outdistance the Soviets once again in micro-electronics is the *raison d'être* of the Pentagon's high-priority, tri-service Very High Speed Integrated Circuits (VHSIC) program. A resounding success so far in development, the program is beginning to turn out semiconductor chips for military computers that should make them capable of processing signals and data at least a hundred times faster than is presently possible....
>
> ...the ATF will be designed for VHSIC data and signal processing right from the start.[3]

VHSIC was structured in two phases, with Phase I equipment based on 16-bit processors and Phase II based on 32-bit processors. The VHSIC program encompassed not only the development of improved processors based on advances in integrated circuit technology, but also the establishment of qualified manufacturing processes, computer-aided design tools, built-in diagnostics, and Ada software programming capabilities. The end result is high reliability, high performance signal and data processing combined with much more compact packaging than previously possible.

As the ATF program got under way, it became apparent that ATF would be the flagship Air Force program for the implementation of integrated avionics. Consequently, Pave Pillar became closely associated with the ATF program. When industry proposals for Pave Pillar were received on September 25, 1984, the bidders were the same seven aircraft prime contractors who had participated in the ATF CDI studies. Pave Pillar at that time was envisioned as a three-phase effort to define a common avionics systems architecture.[4] Funding shortfalls apparently delayed the start of Pave Pillar; the ATF SPO worked with Air Force Wright Aeronautical Laboratories (AFWAL) to accommodate the funding limitations, and Pave Pillar contracts of approximately $1 million each were awarded to all seven of the ATF contractors on September 1, 1985.[5]

The ATF/Pave Pillar integrated avionics work, in concert with an overall DOD-wide push toward integrated electronics, led to the formation of the Joint Integrated Avionics Working Group (JIAWG) in late 1986.

JIAWG was an Army/Navy/Air Force team that established common avionics module specifications. The Air Force and Navy had already been working together for approximately two years to develop Advanced Tactical Aircraft (ATA)/ATF avionics commonality. The JIAWG was formed shortly after the Army joined in this effort with the Light Helicopter Experimental (LHX) program. Each participating service agreed to apply the resulting common architecture and equipment to its respective advanced weapon systems, starting with those already noted.[6]

In July 1987, a software task group of the JIAWG (led by Maj. Bob Lyons of the ATF SPO) met with all six prime contractor teams involved in the ATF, ATA, and LHX programs to discuss software issues. Jon T. Graves, the ATF deputy system program director, noted, "This is the most difficult area of common avionics, but the one with the greatest potential if truly reusable software can be achieved." The JIAWG issued a draft specification for a 16-bit common avionics processor (CAP-16) to industry for comment in August 1987.[7]

In 1988, the Air Force requested its Scientific Advisory Board (SAB) to conduct a review of the ATF avionics program. The board conducted its summer study on integrated avionics in July and produced approximately 50 major findings, which led to various modifications to the program; however, their overall conclusion was that the risks were manageable. In particular, the SAB findings emphasized the need for design discipline and proper documentation to "ensure that a complete and consistent view of requirements is developed and flowed down to all levels of the design organization."[8] Overall, the SAB study recommendations led to an improved understanding and documentation of the integrated avionics architecture and to improved coverage of risk areas. Key members of the study group have subsequently continued to review the ATF avionics program. Successful demonstrations of core prototypes by both ATF teams were beginning to occur by the time the SAB study was completed, and in late 1988 the decision was made to commit to the new integrated avionics architecture that had grown out of Pave Pillar.[8,9]

This commitment supported an overall, planned decision process, encompassing 1) commitment to modular, integrated architecture; 2) identification of specific avionics functions to be included; and 3) definition of functional performance requirements. Although, at the top level, these events occurred in the order listed, important activities in support of the last two were taking place from the beginning of demonstration and validation (Dem/Val). Specifically, the most basic and necessary avionics functions were included in the baseline from the beginning, and effort began as early as appropriate to define performance requirements so that development work could proceed.

In December 1988, the JIAWG released the Common Avionics Baseline IIB (CAB IIB). The key issue addressed in this update was commonality between the Army, Navy, and Air Force requirements for the module set. Work on CAB III began almost immediately after the release of CAB IIB, in early 1989.

JIAWG modules are slightly larger than $6 \times 6 \times 0.5$ in. and fit into cooled, powered slots in racks on the aircraft. A small number of different types perform all of the necessary generic functions—digital data processing, analog-to-digital conversion, bulk memory, power supply, etc. Specific tasks are programmed rather than built in to the hardware. This allows tasking to be assigned and reassigned during a mission, based on the need at any given time, and also to accommodate failures or damage. This reduces the amount duplicate hardware necessary to achieve functional redundancy. All the modules incorporate built-in-test (BIT), and all are programmed in Ada. The cost of a typical module (as of 1990) was around $30,000 but was expected to drop to $10,000 in quantity production.[10] On the F-22A, more than 100 common avionics modules are located on liquid-cooled racks in two bays, one in either side of the fuselage nose section. The modules are individually replaceable, and the bays are accessible from ground level. The hierarchy from VHSIC chips to the common module set to full avionics systems is illustrated in Fig. 67.

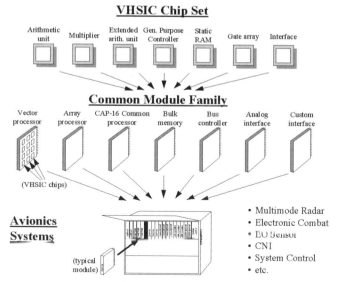

Fig. 67 Hierarchical structure for common integrated avionics.

INEWS started in 1983, following approximately four years of WL-sponsored electronic warfare integration studies. INEWS was managed jointly by WL, ASD/RW (the ASD deputy for Avionics), and the U.S. Navy.[1] In September 1983, concurrently with the beginning of the funded ATF CDI activity, Col. Piccirillo met with Col. Kelly of ASD/RWW (the Electronic Warfare SPO) to formulate the interface between the ATF and INEWS programs. Thereafter the ATF program maintained close contact with the INEWS program.[6] INEWS was ultimately moved into the ATF SPO to ensure maximum transfer of results to the ATF contractor teams.

INEWS encompassed the following functions/features[1]: 1) sensor integration for situational awareness, 2) weapon alert systems, 3) missile alert, 4) countermeasures response, 5) fault tolerant architecture, and 6) modular avionics.

The INEWS concept definition (CD) RFP was released in December 1983 to 33 firms with the capability to manufacture the envisioned electronic warfare (EW) system. Six teams responded, and five were awarded concept definition contracts of approximately $3 million each in July 1984. In June 1986, two teams were awarded contracts of approximately $48 million each for a 31-month Dem/Val phase: TRW/Westinghouse and Sanders/General Electric.[11] (The other three INEWS teams in the CD phase were Hughes/Loral, ITT/Litton, and Raytheon/Northrop.) As of June 1986 the U.S. Air Force was reportedly encouraging these companies to join one of the winning Dem/Val teams in order to "inject some of the better ideas from losing INEWS concepts into the concepts that have survived."[12]

The total cost of the INEWS program was projected at that time to be $8 billion, including the production of 1500 units for ATF and ATA.[13] "INEWS is to be integrated fully with other systems in the ATF and ATA—including sensors, countermeasures devices, and aircraft avionics—to give combat crew members a timely attack warning and automatic countermeasure response." INEWS Dem/Val was expected to last 31 months and be followed by a 60-month full-scale development (FSD) phase, a schedule that was designed to be compatible with the development schedule of the ATF. "One joint venture team will be selected for the full-scale development phase, and members of that final team will compete for production of the system. INEWS production will be in parallel with production of the ATF and ATA."[14]

The decision to fly ATF prototypes during Dem/Val resulted in a period of uncertainty over what impact that would have on the INEWS schedule. There was speculation that INEWS would have to be accelerated to meet the new ATF prototype schedule. At the same time, the prime contractors bidding for the ATF Dem/Val phase reportedly solicited proposals from avionics companies, including the losing INEWS bidders, to develop an EW system for the ATF outside of the INEWS program. In that case, the EW system would be contractor furnished equipment (CFE), an arrangement that the prime contractors apparently considered preferable to a government-furnished INEWS system.[20] However, throughout the ATF Dem/Val phase, INEWS remained the focal point for development of integrated EW technology for the ATF.[1]

During the summer of 1986, immediately following the announcement that ATF prototypes would be flown during Dem/Val, exactly what "prototyping" meant in the ATF program was not yet widely understood.[13,15] In fact, INEWS would not be required on the ATF Dem/Val prototype aircraft. Prototyping actually lengthened the ATF Dem/Val phase and delayed FSD start, thereby providing *more* time to develop the necessary avionics technologies and systems. However, it was felt that an INEWS system might be required, at least in breadboard form, for the avionics testbed portion of the ATF Dem/Val phase. The necessary contract modifications were accomplished in 1987–1988, to ensure that INEWS would achieve the right level of definition during ATF Dem/Val.

In 1988, management of INEWS was transferred to the ATF program. Concurrently, the Seek Spartan program was initiated to transition INEWS technology to other weapons systems including the F-15, F-16, F-111, and A-10.[1]

The Sanders (now Lockheed Sanders)/General Electric team was ultimately selected to proceed with full-scale development of the F-22's electronic warfare suite. The system, based on INEWS technology, includes a radar-warning receiver, missile approach warning, infrared (IR) and radio frequency (RF) countermeasures, and electronic support measures functions. It consists of about 70 Standard Electronics Model E (SEM-E) format modules that fit together with a central integrated processor being developed by Hughes.[16]

INTEGRATED COMMUNICATION, NAVIGATION, AND IDENTIFICATION AVIONICS

ICNIA was a companion program to INEWS in developing integrated avionics technology in specific functional areas to support the ATF. As indicated by the name, ICNIA focused on integration of the systems that perform communication, navigation, and identification (CNI) functions.

Wright Laboratories first began to study the concept of CNI integration during the 1960s, although no major programs were started at that time. From 1978 to 1980, a multifunction/multiband airborne radio program developed four competing integrated CNI architecture definitions. The ICNIA program began in 1981. Wright Laboratories sponsored two competing designs by Texas Instruments and TRW for the first phase, from 1981 to 1985. The Army joined the program early in this phase. Goals were reconfigurable fault tolerant architecture, modular avionics equipment, and two-level maintenance with BIT.

In 1986, TRW won the competition for the second phase, referred to as the advanced development phase. ICNIA was selected for ATF in 1986, and the Navy joined the program in 1987. During ATF Dem/Val, the ICNIA schedule was fine-tuned to support ATF, and the ICNIA advanced development model (ADM) design was changed to be more compatible with Pave Pillar, the overall avionics architectures derived from Pave Pillar. The first ADM ICNIA terminal was expected to be delivered in 1987 or 1988, with several terminals delivered in 1989.

> The ICNIA advanced development model terminals will include prototype software and hardware in brassboard form. According to Gress [Lt. Col. James W. Gress, chief of the joint avionics division within the ATF SPO], the hardware group will include a radio-frequency (RF) group, a signal processing group, and a data processing/communications security (COMSEC) group. He said that the advanced development model hardware and software "will demonstrate the feasibility of ICNIA technology and be used to develop the CNI integration requirements for the ATF."
>
> "The most capable [ICNIA] terminal," Gress said, "will have the following waveform capabilities: JTIDS [Joint Tactical Information Distribution System], TACAN [tactical air navigation], UHF, VHF, GPS [global positioning system], MLS/ILS [microwave landing system/instrument landing system], IFF [identify friend or foe] transponder, and IFF interrogate."[17]

The total investment in ICNIA, through approximately 1991 (corresponding to ATF EMD start), was $126 million, of which the Air Force funded $106 million and the Army and Navy each approximately $10 million.[1] TRW is developing the production CNI system for the F-22.

ICNIA led into a common module program (CMP) late in the ATF Dem/Val phase. CMP was managed by the ATF SPO and was chartered to extend and refine the ICNIA design baseline for greater commonality to meet the CNI needs of all aircraft programs participating in the JIAWG.

Regarding both INEWS and ICNIA, the ATF program office noted at the start of Dem/Val that "these programs are not designing the production avionics configuration for the ATF—they are supporting programs that contribute to the technology base and initial design on which the integrated avionics suites of the ATF contractors are based."[17]

Various congressional actions, intended to place greater emphasis on and/or assume greater control of ATF avionics development, often seemed to overlook this fact. For example, in July 1985 congressional concern over the availability of avionics integration technology (in time for ATF FSD) led to the creation of a separate program element (PE) for this area. Sixty million dollars was put in this new program element. However, this turned out to be available for INEWS and ICNIA only and could not be used for "ATF avionics integration" outside of those specific programs.[18]

Not only would the ATF avionics require further development beyond INEWS and ICNIA within the functional areas corresponding to each of those programs, but also in other functional areas (most notably the radar and other sensors) together with the core avionics that would provide central management of the CNI, EW, and all other avionics systems. In 1985, Col. Piccirillo noted that Pave Pillar was just as essential as the various functional integration efforts because it was developing "the core architecture which is necessary if INEWS/ICNIA and all other advanced avionics systems are to be efficiently integrated" in the ATF.[19]

Other proposed congressional budget language in late 1985 would have prevented the ATF program from even using any of its own funding for ATF radar or electro-optical sensor efforts. This seemed to be based on the idea that the various joint-service avionics programs would eliminate the need for any system-specific avionics development. However, the physical sensors on the ATF—the antennas and other sensing equipment, located all around the F-22, that provide data to the avionics processors—would almost have to be unique to the ATF because none of the other services were attempting to implement integrated avionics on a supersonic, very low observable platform. Col. Piccirillo noted that "a major part of our RFP effort is in the development of radar/EO [electro-optical] sensors for ATF.... Obviously, ATF requires a tailored sensor suite to be effective (*especially* in light of the new RFP amendment)," referring to the more stringent signature goals directed by the Office of the Undersecretary of Defense for Research and Engineering (OUSDR&E) shortly after Dem/Val RFP release.[20] Part of the necessary sensor technology, the Ultra Reliable Radar, is discussed in the following section.

ULTRA RELIABLE RADAR

The central feature of the URR was that it was an "active array" composed of over 1000 transmit-receive (T/R) modules (see Fig. 68). Current generation phased array radars (as found in the B-1B or the F-16) have a phase shifter at each array element, but only a single transmitter and receiver for the whole radar system. In the active array radar, each T/R module contains a transmitter, a phase shifter, a receiver, and a pre-amplifier, all on a single chip. Both types share some important advantages, including the ability to perform extremely rapid beam steering. This allows the radar to track a set of known targets and search for others in a manner which is so rapid that it appears simultaneously to the pilot. Because the steering is accomplished electronically, the physical antenna-pointing hardware is eliminated.

However, the "active array" achieves further dramatic improvements in the following areas[10]:

1) *Performance (i.e., sensitivity)*—Active array avoids noise that is normally generated in the paths from the transmitter to the antenna, and from the antenna back to the receiver, because these paths are now much shorter and simpler (all self-contained within each T/R module).

2) *Versatility*—Because everything from transmission through reception is done at each element, the array can be subdivided into overlapping sub-arrays with different elements doing different things at the same time, for example, to produce multiple beams with different characteristics simultaneously. As information processing capacity advances over the next several decades, this feature will undoubtedly lead to a variety of novel operat-

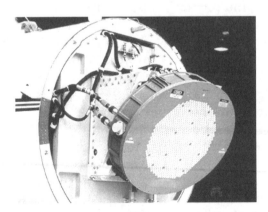

Fig. 68   Prototype active array radar tested in the Northrop avionics flying laboratory.

ing modes and quantum enhancements in electronic counter-countermeasures (ECCM) capability.

3) *Reliability*—In an active array, there is no longer the possibility of a single-point failure at the transmitter or receiver. Several T/R modules could fail without noticeably impairing the performance of the radar. This will greatly reduce the number of maintenance actions required; failed T/R modules will not have to be replaced immediately but can be replaced when it is convenient to do so.

The Ultra Reliable Radar program (1984–1989) represented the culmination of a large number of radar technology programs from the 1960s through the 1980s. These included the following:

1) Forward-looking advanced multimode radar (FLAMR)—1965–1974. Technology subsequently applied to F-15, F-15E, F/A-18, and TR-1.
2) Solid-state phased array technology base—pre-1965 to 1980.
3) Electronically agile radar—1970–1980. Used on F-16C/D, B-1B.
4) Galium Arsenide (GaAs) technology base—1970–1988. Approximately $21 million WL investment in devices, $20 million in circuits and components, and $37 million in modules and arrays (with matching contractor funding). GaAs overcame key limitations of earlier silicon-based technology.
5) Electronic counter-counter measures (ECCM)—1977–1990.
6) Solid-state phased array demonstration—1981–1984. Demonstrated performance equivalent to the F-15 radar, excellent antenna patterns, 2500 h array-level MTBF, and two-level maintenance capability. Acquisition cost was the major outstanding issue.
7) Work continuing after 1990 (i.e., in conjunction with the EMD phase of the ATF) focuses on achieving lower cost, higher reliability, and improved power management for enhanced flexibility and survivability [i.e., low probability of intercept (LPI) modes].[1]

In 1987, Hughes/General Electric and Westinghouse/Texas Instruments competed for the development of the ATF radars, to be based on URR technology. The sources for the radars, unlike the EW and CNI suites, were selected by each ATF prime contractor. Both competing ATF teams selected the Westinghouse/Texas Instruments team; this team is now responsible for development of the AN/APG-77 radar for the production F-22 as well as the electro-optical (EO) sensors.

JOINT TACTICAL INFORMATION DISTRIBUTION SYSTEM

The Joint Tactical Information Distribution System (JTIDS) program was initiated in 1975 to provide a "secure, jam-resistant digital information link." The Air Force was named by the Office of the Secretary of Defense (OSD) as the lead service. In the late 1970s, it became apparent that the JTIDS terminal would be too large for fighter aircraft and also for mobile Army units. Development of the smaller, lighter JTIDS class 2 terminal was initiated in the late 1970s. Two variants evolved: the class 2 and the class 2H, which would have a high-power amplifier for greater range than the basic class 2 terminal. In the mid-1980s an Army requirement was defined for an even smaller version than the basic class 2. This resulted in the class 2M, which was put on a separate development track.[21]

In 1986–1987, class 2 JTIDS testing was performed on the F-15. The system demonstrated an MTBF of 17 h, compared to a requirement of 102 h that was established in a 1981 SECDEF decision memorandum. It performed at or above threshold in all other areas. Approval for low rate initial production (LRIP) of the 2 and 2H was given in October 1989. Three one-year LRIP lots were planned, but final approval of the second and third lots were subject to meeting certain criteria including improved reliability. A full production decision was planned for late 1993, following Navy evaluation, Air Force multiplatform operational test and evaluation (OT&E) in March 1993, and Army initial OT&E (IOT&E).[21]

Estimated JTIDS program costs as of 1990 were $2 billion for development and $1.9 billion for production. Development was expected to be completed in 1995. The total production run was expected to be 1441. The Air Force originally planned to equip approximately 140 F-15s with JTIDS, but in the late 1980s reduced the number to 20.[21]

JTIDS interoperability was a requirement for the ATF from the outset (i.e., stated in the 1984 Statement of Operational Need). The requirement was deleted in the April 1990 requirements update and reinstated (receive-only) in 1996. Thus, the present requirement is that the F-22 will be equipped with JTIDS receivers, but will not be capable of JTIDS transmission.[22]

COCKPIT

The F-22 cockpit will make extensive use of flat-panel multifunction displays. The YF-22A cockpit had utilized liquid crystal multifunction displays (MFDs) with finger-on-glass (FOG) control. "Difficulties with the FOG system, due primarily to a lack of tactile feedback, have led to a decision to return to a conventional push button switch system"

around the display screens, for the F-22.[16] (This was the same reason
that the nonmoving side stick controller of the YF-16 was replaced by
a moving side stick on the production F-16.[23])

In the case of the F-22 cockpit displays, active-matrix liquid crystal
display (AMLCD) technology offers a significant advantage in weight,
volume, and cost along with major improvements in reliability. It also
provides better contrast and brightness over its passive matrix predecessors. As Air Force Lt. Col. Charles Pinney (who was assigned to the
SPO during the latter stages of the Dem/Val program and on into the
early part of EMD) recently recalled

> Developing a rugged, producible AMLCD version capable of
> withstanding the demanding fighter aircraft environment (temper-
> ature extremes, shock, and vibration) and the tight display stan-
> dards (minimum failure of pixels per unit area) was a formidable
> task. The Japanese were leaders in this industry. A noted Japanese
> manufacturer (Hoshidan) decided not to pursue the unique F-22
> military application. Also, many felt that the F-22 supplier should
> be a U.S. company. Eventually, one U.S. company, Optical Imaging
> Systems of Troy, Michigan became the vendor. As this small busi-
> ness grew to meet eventual F-22 production needs, it eventually be-
> came the source for AMLCDs for other U.S. weapons systems in
> development or undergoing modernization upgrades.

WEAPONS TECHNOLOGY

COMPRESSED CARRIAGE AIM-120 ADVANCED MEDIUM RANGE AIR-TO-AIR MISSILE

Internal weapons load has always been one of the central tradeoffs
in the ATF design. Because the ATF must be sized to carry its payload
internally, the number of missiles must be carefully balanced against
aircraft size and weight. When the ATF first became focused on the air
superiority mission, six was considered to be the minimum acceptable
number of missiles with eight or more highly desirable. The determination was also made that the ATF must not be dependent upon new
weapons to achieve its basic design mission capability.

Early in the program, the following compromise was selected: the
ATF would be capable of carrying six missiles of existing design [a
mix of AIM-120A/B advanced medium range air-to-air missiles
(AMRAAM) and AIM-9 Sidewinders] or eight missiles of new or
modified design for compressed carriage.

The Air Force Systems Command's Armament Division contracted
with Hughes for a folding-fin AMRAAM feasibility study in September 1984. The ATF SPO provided $200,000 for this effort. Mid-term re-

sults, briefed to both the AMRAAM and ATF SPOs in March 1985, were favorable and indicated that "internal carriage and launch are feasible and several options are possible."[24] The options included folding fins, and/or fins of reduced size, together with a variety of techniques for ejection and launch from the internal bays. Some of the concepts evaluated in the Hughes feasibility study are illustrated in Fig. 69.

Following the folding-fin AMRAAM feasibility study, a compressed carriage demonstration program was formulated. A branch within the projects division of the AMRAAM SPO was formed to work compressed carriage efforts for ATF. A sources-sought synopsis to "Analyze, Design, Manufacture and Demonstrate Advanced Missile Carriage and Release Techniques" for ATF application was coordinated with the ATF SPO and was published on March 8, 1985. Responses were due on April 7.[24]

In November 1985, the Armament Division was preparing for a November 15 RFP release for the demonstration, which was planned to include analysis, wind-tunnel test, and launch/separation from an internal bay at representative flight conditions prior to the start of ATF FSD. However, in early November the AMRAAM compressed carriage demonstration (ACCD) was placed on hold:

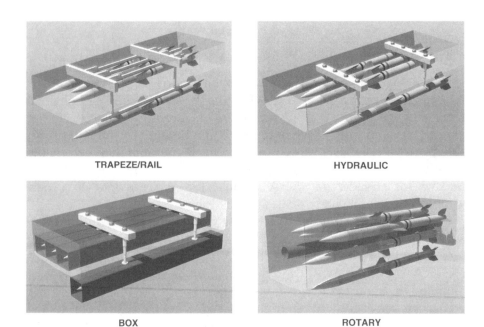

Fig. 69   Early ATF compressed missile carriage concepts.

The ACCD was placed on hold due to the intense political issues surrounding the AMRAAM program at this time. The Air Staff is concerned that any effort, even a technical feasibility demonstration, dealing with the need for eventual modification to the baseline AMRAAM (i.e., folding fins or redesigned fins for ATF usage) could be detrimental to the AMRAAM program in Congress. Therefore, the Air Staff has decided to delay the ACCD effort until after the Secretary of Defense's certification of AMRAAM. This certification is required by the FY86 Authorization Bill and must be accomplished prior to 1 Mar 86.[25]

The AMRAAM program was going through a major review and restructuring at that time because of difficulties in its development. Even after the referenced SECDEF certification was granted for the overall AMRAAM program, the ACCD continued to be delayed.

Following the restructuring, the AIM-120 AMRAAM eventually completed its development and entered service in 1991. The AIM-120C (Fig. 70) has been introduced as a preplanned product improvement (P3I) and features reduced-span fins for internal carriage in the F-22. Deliveries of the AIM-120C began in 1996. The AIM-120C also includes improved lethality and electronic counter-countermeasures (ECCM) capabilities.

SHORT-RANGE MISSILE DEVELOPMENT

Initially two possible short-range missiles were considered for the ATF: the multinational advanced short range air-to-air missile (ASRAAM), or a version of the AIM-9. The memorandum of understanding among the countries participating in ASRAAM essentially collapsed, however, leaving the AIM-9 as the remaining choice.

The preferred version will be the AIM-9X with improved off-boresight capability (when used with a helmet-mounted sight); thrust vectoring and a high-angle-of-attack airframe design for improved maneuverability; and a seeker based on focal plane array technology rather than the reticule-type seekers used on the current AIM-9s. The

Fig. 70   The AIM-120C features reduced-span fins. It was intended for internal carriage with the F-22 and can also be employed from external carriage systems on other fighter aircraft.

new seeker offers better resolution and improved infrared counter-countermeasures (IRCCM) capability (i.e., will be harder to deceive with flares, etc.). One of the key advantages of the AIM-9X over earlier versions of the AIM-9 is that it will be delivered as a total package with the Joint Helmet Mounted Cueing System (JHMCS) to provide the off-boresight capability.

Although the AIM-9X has been the subject of informal concept studies since the 1980s, it only became a formal acquisition program in 1993. AIM-9X is a joint program with the U.S. Navy as the lead service. Hughes and Raytheon participated in a competitive Dem/Val phase, and Hughes was selected for EMD that began in January 1997. Hughes's seeker design for the AIM-9X is based on work that Hughes did for the ASRAAM program. Aerodynamic features of the AIM-9X are derived from the U.S. "Box Office" and "Boa" programs, which involved several demonstration launches of new airframe designs. IOC of the AIM-9X is expected in 2002. This actually places the United States somewhat behind in short-range missile capabilities at the present time, since both Israel and Russia already have advanced IR missiles (the Python and the Vympel R-73, respectively) with capabilities that exceed those of the current-generation AIM-9s in some respects.[26]

STRUCTURES AND MATERIALS

Wright Laboratories and the airframe and engine contractors have developed a variety of advanced materials technologies. Laboratory programs in the late 1970s through the 1980s were primarily in the three areas of materials for gas turbine engines, composite materials for airframes, and advanced metallic structures. Both the Materials Laboratory and the Flight Dynamics Laboratory (AFFDL) have been heavily involved in this area, with the flight dynamics work concentrating on flight qualification, demonstration of composite airframe components, and applications such as ceramics and coatings for two-dimensional nozzles.

MATERIALS FOR GAS TURBINE ENGINES

Processes for manufacturing engine components evolved through several stages during the period leading up to the ATF/F-22 program. Directionally solidified turbine blades were developed from the late 1970s to mid-1980s, followed by single-crystal techniques in the mid-1980s. Forging processes proceeded from powder metallurgy (which was applied to the F100/F110 generation of fighter aircraft engines) to techniques that permitted single-piece "integrally bladed rotors" to be fabricated (i.e., the entire rotor with all of its blades is one piece, rather than having each blade mechanically connected to the ring). Titanium 6-2-4-2 alloy was used on fan blades in the late 1970s, and by the late 1980s hollow fan blades were being fabricated. Advanced metallurgical capabilities such as rapid solidification rate (RSR) facilitated reduced stages for equivalent amounts of work. In the RSR process alloys are frozen in solution by cooling the molten metal at rates of 1 million °F/s. The use of RSR alloys allowed higher rotational speeds for more work per stage and, consequently, fewer parts required; RSR parts were also capable of operating at higher temperatures for equivalent life. RSR was originally planned for introduction in the F119 production engines. However, improvements in turbine casting techniques allowed intricate cooling passages to be formed inside turbine blades, permitting cooler blade operation and reducing the required blade strength. This was found to be a cheaper manufacturing process, so RSR was never implemented.[27]

Other programs addressed damage tolerance and diagnostics, composites on certain cold-section parts such as front frames, thermal barrier coatings, lubricants, and radar absorbent materials. Many of these technologies were applied to the ATF engines, sometimes after further proprietary development by the engine contractors.

AIRFRAME MATERIALS TECHNOLOGY

Composite materials used on the YF-22A and F-22 include wet and dry thermoplastics, Bismaleimide (BMI) thermoset, and epoxy thermoset materials. Thermoplastic and thermoset materials and processes were developed throughout the 1980s. (The distinction between thermoset and thermoplastic materials is that thermosets, once cured, cannot be reprocessed, whereas thermoplastics can be reshaped by reheating them. This quality may be able to provide some improved battle damage repair possibilities.[16,28])

Advanced metallic structural technology developed since the late 1970s includes elevated temperature aluminum, aluminum–lithium alloy, metal matrix composites, and processes for improved cost and producibility of titanium. Elevated temperature aluminum materials were developed from 1978 to 1982. From 1982 to 1985, processing techniques were developed, and the materials began to be applied to aircraft structures in the 1980s. Aluminum–lithium was used on the wing skins of the F-15 STOL and maneuver technology demonstrator in the mid-1980s. Metal matrix composites were demonstrated on a horizontal stabilizer and various supplemental structural components during the 1980s. Wright Laboratory's built-up low-cost advanced titanium structure (BLATS) program of the late 1970s led to techniques that were applied on both teams' ATF prototypes, as well as on the F-15E. Many other advances in structures, materials, and manufacturing processes have been applied to the F-22. Some have taken place gradually and incrementally, and therefore cannot be linked with specific programs.

However, the use of composites on the ATF did not meet many early expectations. This has been a recurring theme in the development of advanced aircraft. Predictions of 50% or higher use of composites, leading to dramatically lighter airframes, often turn out to be optimistic. According to YF-22A team General Manager Sherm Mullin, "I had one major technical disappointment: Our failure to successfully apply as much composites to the primary structure of the airplane as originally planned. Our stated objective was to utilize composites for 40% of the aircraft structure by weight and to achieve a structural weight reduction of 25% or better compared to aluminum...." The actual composites usage on the YF-22A was only 23%, as shown in the list that follows. "We did achieve our goal of a 25% weight reduction compared to aluminum" for those areas where composites were used.[29] Here is the YF-22A aircraft's structural composition[28]:

33% aluminum
2% advanced aluminum alloys
24% titanium
5% steel
13% graphite thermoplastics
10% thermosets
13% other miscellaneous

Thus, metals (steel, aluminums, and titanium) accounted for 64% of the structure, composites accounted for 23%, and other miscellaneous materials 13%. The F-22 structural materials, as of early 1992, were planned to be as follows:

11% aluminum
33% titanium (22% Ti 62222, 11% Ti 6-4)
5% steel
15% graphite thermoplastics
20% thermosets
16% other miscellaneous

The corresponding subtotals are 49% metals, 35% composites, and 16% other miscellaneous. The increased use of titanium (33% on F-22 vs 24% on YF-22) is one reason for cost growth relative to earlier estimates. More recent sources indicate a further increase in the percentage of titanium, accompanied by reductions in the use of composites.[30]

Low Observables

Obviously, many advances in low-observables technology have been made in support of the ATF program over the past 10 years. Initially, the technologies developed in and demonstrated in the Have Blue and F-117A programs (see Appendix) provided the main source for Lockheed's initial ATF conceptual studies in the early 1980s. Major investments in other low-observable aircraft developments in the years since the F-117A was fielded, as well as in the ATF/F-22 program itself, have provided a rich technology base in this area. These investments have led to significant advances in overall design capability and testing methodologies, radar absorbent materials and structures, low-observable sensors and avionics, low-observable air data sensors, engine treatments, and low-observable inlet/exhaust installations. For example, as noted by Lt. Col. Pinney

> The radome represented one of the most complicated design features on the aircraft; it affected drag, inlet performance, LO/RCS and radar performance. The F-22 wideband radome was originally novel and high risk, especially from a manufacturing perspective. The design at the end of Dem/Val consisted of drilling hundreds of small holes and filling them with plugs to provide the correct antenna patterns and coverage across the frequencies of interest. The breakthrough in EMD involved etching the surface of the frequency selective substrate. While this technique might seem relatively straightforward, the complex chined shape of the radome posed a geometric nightmare to achieve the appropriate RF and LO performance all the way around.

Because of the highly classified nature of this technology, it is not possible to address the details of most of these technologies.

Supporting Engine Technology Programs

Advances in military turbine engine technology since the introduction of the jet engine have primarily been developed through joint government/industry-funded research programs. Applied research on engine components has been instrumental in achieving ever increasing thrust-to-weight ratios, improved durability, and better operability characteristics. Initial evaluation and/or validation of component technologies in Wright Laboratory's Compressor Research Facility (CRF) and Turbine Research Facility (TRF), as well as other government and industry facilities for other components and materials testing, has been essential in developing new levels of performance.

The Advanced Turbine Engine Gas Generator (ATEGG) program began in the late 1960s as a way to integrate these advanced components into an engine core (the gas generator)—the compressor, combustor, and high-pressure turbine, which are all connected by the innermost shaft (the high-pressure spool). After a core component was successfully demonstrated in rig tests, it was integrated into an ATEGG demonstrator for in situ evaluation. These critical tests assessed the performance and mechanical characteristics of the component, interacting with other components in a realistic engine environment.

The Aircraft Propulsion Subsystem Integration (APSI) program concentrated on the other engine components—the fan and low-pressure turbine, which are connected by the outer shaft (the low-pressure spool), as well as controls, mechanical systems, exhaust systems, and nozzles. ATEGG and APSI were integrated through testing on a Joint Technology Demonstrator Engine (JTDE), which began in the mid-1970s.

In the late 1980s, the industry/government engine technology community set a number of ambitious goals and integrated these and other efforts as the Integrated High Performance Turbine Engine Technology (IHPTET) program. The primary goal is to double the thrust-to-weight ratio of fighter/attack engines (compared to the ATF Dem/Val engines) early in the next century. A three-phase program was set up to accomplish this effort, and Phase II goals (which include a 60% thrust-to-weight increase) are now being met.

FLIGHT DEMONSTRATOR AIRCRAFT PROGRAMS RELATED TO ADVANCED TACTICAL FIGHTER

In addition to programs in specific technology areas, many advanced technology flight demonstrator programs were promoted as supporting the Advanced Tactical Fighter. These are discussed in the following sections. In some cases, their link to ATF was purely promotional, and the demonstrations contributed little to the technology that was actually used on the ATF. In other cases the programs did address key technology needs for the ATF. The F-15 STOL and Maneuver Technology Demonstrator (S/MTD) is the best example. This program provided valuable experience in the use of two-dimensional vectoring/reversing nozzles, integrated flight/propulsion control, and also contributed to cost savings in the ATF program through some important requirements adjustments (involving the deletion of thrust reversing).

ADVANCED FIGHTER TECHNOLOGY INTEGRATION

As noted earlier, the Advanced Fighter Technology Integration (AFTI) program was initiated in 1973 by the Air Force Flight Dynamics Laboratory and was expected to be supportive of the ATF. The AFTI plan was to sponsor some initial studies and then select one or several companies to build demonstrator aircraft. Candidate technologies included vectored thrust, circulation control/enhancement for increased lift, direct lift/direct sideforce control, fly-by-wire, relaxed static stability, maneuver load control, flutter suppression, in-flight thrust reversal, and electronic integrated flight/propulsion control.

The preliminary studies in 1974–1975 led to a three-phased plan for AFTI. The first phase (AFTI-I) would focus on flight and fire control technologies; AFTI-II would demonstrate wing technologies for multiple flight regimes; and AFTI-III would integrate the results of the first two phases.[31]

Two contracts were awarded under AFTI-I in 1978: one to McDonnell Douglas to modify an F-15 and one to General Dynamics to modify an F-16, as flight/fire control technology testbeds. The AFTI F-15 featured modified flight control laws. It flew in 1982 and successfully demonstrated the use of integrated flight/fire control techniques for improved air-to-air gunnery and maneuvering air-to-ground bomb deliveries.[31]

The AFTI F-16 (Fig. 71) was more extensively modified. It had a pair of near-vertical control surfaces mounted on the engine inlet duct. These provided the capability to point the nose to one side or the other with-

Fig. 71   The highly modified General Dynamics AFTI F-16.

out changing the aircraft's flight direction, or conversely, to translate laterally without having to point the nose. Under an earlier (1976–1977) AFFDL program, the F-16 control configured vehicle (F-16 CCV), the first YF-16 prototype had been modified with similar "chin"-mounted control surfaces and new control laws. The F-16 CCV had achieved some success but was limited by the YF-16's analog flight control system. The AFTI F-16 built upon the results of the CCV tests and utilized a triplex digital flight control computer to provide a variety of pilot-selectable control modes tailored to various mission-oriented tasks. Flight tests in 1982 and 1983 concentrated on developing the flight control laws. From 1984 to 1987, a Westinghouse sensor/tracker system was integrated with the flight control system to provide integrated flight/fire control for precision dynamic weapon delivery. The AFTI F-16 has since been used for additional demonstrations. From 1988 to 1992, communication links and other mission subsystems for close air support were demonstrated, and, as of early 1997, the AFTI F-16 is being modified to test electric power generation/distribution and electric flight control actuation technologies in support of the Joint Strike Fighter program.[31–34]

For AFTI-II, Boeing modified a General Dynamics F-111 with a smooth, flexible-skinned variable-camber wing (replacing discreet control surfaces). The F-111 Mission Adaptive Wing aircraft, seen in Fig. 72, flew from 1985 to 1988. The first 26 flights evaluated the aerodynamic properties of the wing as a function of camber and twist. The remaining 33 flights demonstrated automatic control of the wing shape in five different modes: cruise camber, maneuver camber, maneuver load control, gust alleviation, and maneuver enhancement. However, the smooth-skinned wing proved difficult to construct and offered only a small improvement in efficiency relative to the much simpler, conventional hinged control surfaces.[32]

Fig. 72　The Boeing F-111 Mission Adaptive Wing testbed.

AFTI-III, which was intended to apply the flight/fire control technology of AFTI-I and the wing technology of AFTI-II in a single aircraft (probably an all-new aircraft designed from the ground up to exploit these technologies), was never executed.

HIGHLY MANEUVERABLE ADVANCED TECHNOLOGY

The joint NASA/USAF Highly Maneuverable Advanced Technology (HiMAT) program was initiated in support of AFTI in 1975. HiMAT was intended to demonstrate technologies for enhanced maneuverability. Rockwell designed and built two unmanned HiMAT vehicles (one of which is seen in flight in Fig. 73). They were tested from 1979 to 1983. The vehicles were designed with modular features to permit the evaluation of a variety of aerodynamic configurations; however, no such additional configurations were ever tested.[31,35]

COMBAT AIRCRAFT PROTOTYPE PROGRAM

In a program decision memorandum of July 1980, OSD directed the Air Force to initiate the Combat Aircraft Prototype (CAP) program starting in fiscal year 1982. In this program a new prototype project would be started every two years, and each one would have a duration of three years. A streamlined, cooperative approach with industry was envisioned to foster innovation very similar to the "experimental prototyping" initiative of the early 1970s, which had spawned the Lightweight Fighter program. The objectives of CAP were

to provide actual flight demonstration of promising new technologies in order to mature these technologies sufficiently so that

Fig. 73 The Rockwell HiMAT remotely piloted test vehicle.

they will be acceptable candidates for application in the full scale
development of new aircraft…. Full-scale flight testing is needed to
confirm theoretical analyses and subscale verification…. Furthermore, during the current unprecedented gap in U.S. fighter developments, industry design teams have not been afforded the essential experience involved in translating design concepts into actual
flight hardware. The CAP program provides a means to address
these problems and to insure that fighter design capability, technical options and manufacturing processes are available for future
fighter development programs.[36]

ATF was specifically identified as a program that would benefit from
CAP. The program management directive (PMD) for the CAP program[36] noted that pertinent technology demonstrations needed to proceed in parallel with the mission requirements development for the
ATF in order to support the planned 1986–1987 full-scale development (FSD) start and a mid-1990s initial operational capability (IOC).
The General Dynamics F-16XL and the Defense Advanced Research
Projects Agency-(DARPA-) supported Grumman forward-swept-
wing demonstrator were considered likely candidates for the first CAP
project.[37] However, the CAP program was zero-funded during fiscal
year 1983 Air Force budget estimate submission (BES) deliberations
in late 1981.[38] [Nevertheless, both the F-16XL and the forward-swept-
wing demonstrator (X-29) projects proceeded with other sources of
funding and are described later in this chapter.]

CRITICAL TECHNOLOGY DEMONSTRATION

The November 1983 PMD for the ATF program added a third project (the first two were ATF and JAFE), critical technology demonstration (CTD). The critical technology demonstration effort was conducted from 1983 to 1985 by the AFFDL in conjunction with the ATF
SPO to support the ATF concept development investigation by accomplishing the following[1,6]:

1) Identify technologies based on ATF contractor preliminary designs.
2) Plan the development and demonstration (both ground and
   flight) of "those technologies requiring additional development to
   bring them to an acceptable risk level for application to ATF during full-scale development (FSD)."[1]
3) Conduct pre-design studies of technology flight demonstrator aircraft to accomplish the flight demonstration portion of the afore-
   mentioned plan.

Total funding for CTD was approximately $2 million, primarily provided by AFFDL, with the remainder coming from the Air Force Avionics Laboratory (AFAL), the Air Force Armament Laboratory (AFATL), and the ATF program. The effort was conducted via contracts with all seven ATF CDI contractors. The philosophy was to produce a comprehensive technology plan, maximizing the total benefit that could be derived from the companies' internal research and development (IR&D) programs together with contracted research and the government laboratory programs.[1]

It was originally envisioned that a supersonic cruise demonstrator aircraft would be (one of) the demonstrator(s) built as a result of this program. Such aircraft would not be prototypes of specific ATF designs, but rather would be lower cost aircraft specifically aimed at those technologies requiring flight demonstration. However, the decision in 1986 to build ATF prototypes during the demonstration and validation (Dem/Val) phase largely eliminated the need for technology demonstrator aircraft apart from the prototypes themselves. Nevertheless, the CTD effort made important contributions to the overall ATF Dem/Val technology plan.[1,39]

F-16XL

General Dynamics designed and built two examples of the F-16XL, a stretched F-16 featuring a cranked-arrow wing. Initial investigations of this configuration had been conducted in conjunction with NASA starting in 1976. The changes resulted in improved supersonic maneuvering, supersonic cruise capability, increased maximum lift, increased internal fuel capacity, and increased weapons load, relative to a basic F-16. Weapons could be carried conformally, greatly reducing the drag. The effort was primarily company-funded, although the Air Force provided F-16 airframe components. The F-16XL first flew in July 1982. The effort was variously billed as a technology demonstrator (Supersonic Cruise And Maneuver Prototype—SCAMP), or later as a prototype for a proposed multimission F-16 derivative (F-16E). Thus, the F-16XL at various times was seen as supporting the ATF program by demonstrating key technologies and supersonic cruise capability or as competing with the ATF by offering a much cheaper alternative; as an example, in June 1981, the Air Staff diverted $618 million out-year funding from ATF toward development of an operational fighter based on the F-16XL, although this development never actually took place. The F-16XL is seen in flight in Fig. 74.

Fig. 74   The General Dynamics F-16XL.

X-29

The X-29 grew out of several years of forward-swept wing conceptual studies by Grumman (including their unsuccessful bid for the HiMAT program). Many of Grumman's conceptual designs in the early stages of the ATF program also featured forward-swept wings. The forward-swept wing was claimed to offer advantages in maneuverability and efficiency. As just noted, the X-29 was considered a likely candidate for the Combat Aircraft Prototype program before that program was canceled in 1981. Nevertheless, two X-29 aircraft were eventually built under DARPA funding. They began flight tests in December 1984. Although the X-29, seen in Fig. 75, represented an impressive achievement in aeroelastic tailoring, flutter/divergence suppression, control of an aircraft with an unprecedented degree of static instability, and over-

Fig. 75   The Grumman X-29 forward sweep technology demonstrator aircraft.

all technology integration, it did not demonstrate any clear advantage of the forward-swept wing relative to a conventional aft-swept wing.[31,32,37]

SHORT TAKEOFF AND LANDING AND MANEUVER TECHNOLOGY DEMONSTRATOR

The F-15 STOL/Maneuver Technology Demonstrator (S/MTD) used pitch-axis thrust vectoring, canards, integrated flight/propulsion control, and associated pilot–vehicle interface advances for improved maneuverability and short field performance together with carefree handling qualities. It made its first flight in September 1988. The S/MTD was on the original 1983 list of Wright Laboratories technology programs that were of particular importance to the ATF, and thereafter the S/MTD program proceeded in close cooperation with the ATF SPO.[32]

The need for a flight demonstration of two-dimensional thrust vectoring nozzles was identified in 1983, in conjunction with the start of the Joint Advanced Fighter Engine program. The RFP was issued in September 1983 for a testbed aircraft to fly by 1987 with a two-dimensional vectoring and reversing exhaust nozzle (see Fig. 76) and integrated digital flight controls. A key demonstration goal was to demonstrate the use of these features to achieve STOL capability, specifically the ability to operate from a 50 × 1500 ft runway under adverse conditions. General Dynamics and McDonnell Douglas responded to the RFP with proposals for a modified F-16 and a modified F-15, respectively. McDonnell Douglas was selected for the S/MTD effort in October 1984.

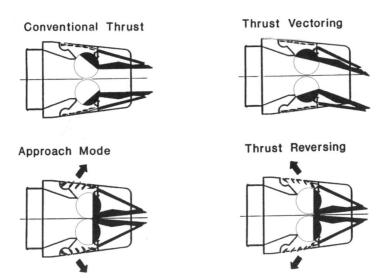

Fig. 76   The various modes for the S/MTD nozzles.

The F-15 airframe was modified with two-dimensional vectoring/reversing nozzles, all-moving canard surfaces (F/A-18A elevators were used), integrated digital flight/propulsion control, advanced pilot–vehicle interface, rough field/high sink-rate landing gear, and advanced approach guidance avionics.

Before the S/MTD began flying, nozzle tests indicated that the structural and thermal demands of thrust vectoring and reversing were greater than anticipated. A redesign of the nozzles was undertaken. Vectoring requirements were relaxed, but even so the weight of the nozzles was considerable. The thermal conditions produced by thrust reversing at high power required a heavy and complex cooling system for the nozzles. This experience led to the deletion of the requirement for thrust reversing in the ATF program in order to save cost and weight. Flight tests of the F-15 S/MTD (Fig. 77), conducted from 1988 to 1991, provided additional experience with several important technologies that would eventually be applied to the F-22.

SUMMARY

The F-22 contains advances in virtually every area of aerospace technology. Many of the technologies were developed specifically for the ATF, and many simply represent advances in the state of the art that have occurred in the time prior to and during the early stages of the

Fig. 77   The McDonnell Douglas F-15 S/MTD aircraft.

ATF program. The Air Force, other government agencies, the airframe and engine prime contractors, and subcontractors and vendors have all contributed to the technology base that makes the F-22 possible. Most of the technology was developed over an extended period of time, starting with basic research and often leading to flight demonstrations on testbed aircraft. The preceding descriptions are not an exhaustive list of all the technologies developed and applied to the F-22, but are only intended to present some of the most prominent and to indicate the timescales and the relationship between technology development and the ATF program. A timeline showing the most significant related technology projects (including propulsion), relative to the overall ATF program, is presented in Fig. 78.

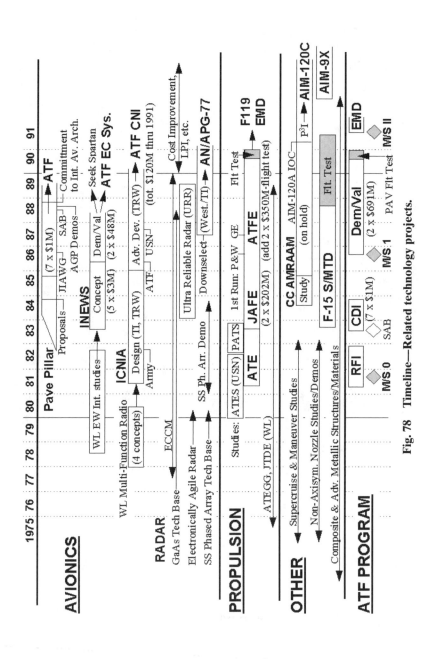

Fig. 78 Timeline—Related technology projects.

REFERENCES

[1]Patterson, R., Capt. USAF, *Wright Laboratory's Role/History in the Advanced Tactical Fighter Technology Development,* Wright Lab. Rept. No. WRDC-TM-90-603-TXT, Wright–Patterson AFB, OH, Jan. 1990.

[2]Dornheim, M. A., "Lockheed Team Will Test ATF Cockpit in Boeing 757 Flying Laboratory," *Aviation Week and Space Technology,* Nov. 10, 1986.

[3]Canan, J. W., "Toward the Totally Integrated Airplane," *Air Force Magazine,* Jan. 1984, pp. 34–41.

[4]Various Weekly Activity Reports (WARs), from ATF System Program Office to the Commander, Aeronautical Systems Div., Sept.–Oct. 1984.

[5]WAR, Sept. 12–18, 1985.

[6]Ferguson, P. C., *Advanced Tactical Fighter/F-22 Annotated Chronology* (Draft), Office of History, Aeronautical Systems Center, Wright–Patterson AFB, OH, Aug. 1996.

[7]WARs, July 6–10, 1987, and subsequent.

[8]Borky, M. J., Col. USAF, *Press Release on Avionics for the Advanced Tactical Fighter (ATF),* ASD 90-2347, Aug. 23, 1990.

[9]Twigg, J. L., *To Fly and Fight: Norms, Institutions, and Fighter Aircraft Procurement in the United States, Russia, and Japan,* Ph.D. Dissertation, Dept. of Political Science, MIT, Cambridge, MA, Sept. 1994.

[10]Sweetman, B., *YF-22 and YF-23 Advanced Tactical Fighters: Stealth, Speed and Agility for Air Superiority,* Motorbooks International, Osceola, WI, 1991.

[11]"Getting INEWS Off the Ground," *Defense Electronics,* Oct. 1986, pp. 46–51.

[12]"Air Force Wants More INEWS Teaming," *Aerospace Daily,* June 26, 1986, p. 489.

[13]"New Advanced Fighter Blueprint Throws Key Schedules out of Kilter," *Defense News,* June 23, 1986, pp. 3, 4.

[14]USAF Press Release, Integrated Electronic Warfare System (INEWS), June 27, 1986.

[15]"DOD Will Accelerate INEWS Program to Meet New ATF Prototype Schedule," *Inside the Pentagon,* Aug. 15, 1986, pp. 5, 6.

[16]Abrams, R., and Miller, J., *Lockheed (General Dynamics/Boeing) F-22,* Aerofax Extra, No. 5, 1992.

[17]Rawles, J. W., "ATF Program Gathers Momentum," *Defense Electronics,* July 1988, pp. 43–51.

[18]Various WARs, July 1985.

[19]WAR, Aug. 15–21, 1985.

[20]WAR, Dec. 12–18, 1985.

[21]*Defense Acquisition Programs: Status of Selected Programs,* General Accounting Office, GAO/NSIAD-90-159, June 1990.

[22]Operational Requirements Document (ORD) Attachment: *Requirements Correlation Matrix* SECRET. (Unclassified information only used from this source.)

[23]Aronstein, D. C., and Piccirillo, A. C., *The Lightweight Fighter Program: A Successful Approach to Fighter Technology Transition,* AIAA, Reston, VA, 1997.

[24]WAR, March 21–27, 1985.

[25]WAR, Nov. 7–13, 1985.

[26]Huber, A., Maj. USAF, AIM-9X Program Element Monitor (PEM), SAF/AQPF, and Brooks, J., ANSER, private communication, June 1997.

[27]"The Core of Navy Air Power," Pratt and Whitney Rept., West Palm Beach, FL, June 1978.

[28]Abrams, R. (Director of Flight Test, Lockheed Advanced Development Company), *YF-22A Prototype Advanced Tactical Fighter Demonstration/Validation Flight Test Program Overview,* 1991.

[29]Mullin, S. N., "The Evolution of the F-22 Advanced Tactical Fighter," Wright Brothers Lecture, AIAA Paper 92-4188, Washington, DC, Aug. 1992.

[30]*Jane's All The World's Aircraft,* 1996–1997 ed., Jane's Information Group, Surrey, UK, 1996.

[31]Lyons, R. R., Maj. USAF, *The Search for an Advanced Fighter: A History from the XF-108 to the Advanced Tactical Fighter,* Air Command and Staff College, Air Univ., Maxwell AFB, AL, April 1986.

[32]Aronstein, D. C., and Piccirillo, A. C., *Comments and Observations on Flight Demonstrator Programs,* ANSER, Arlington, VA, Oct. 25, 1994.

[33]Smith, R. A., Lt. Col. USAF, Joint Strike Fighter Flight Systems Integrated Product Team Lead, private communication, Jan. 1997.

[34]*Jane's All The World's Aircraft,* 1977–1978 ed., Jane's Information Group, Surrey, UK, 1977.

[35]*Jane's All The World's Aircraft,* 1983–1984 ed., Jane's Information Group, Surrey, UK, 1983.

[36]Program Management Directive No. R-Q R-Q 1057(1)/63242F, *Combat Aircraft Prototype (CAP),* USAF/RDQT, April 14, 1981.

[37]Adams, J., Col. USAF, (TAC/DRD), to TAC/CC, CV3, XP, Staff Summary Sheet, Subject: *Future Fighter Alternatives Study Steering Committee Meeting,* Nov. 3, 1980.

[38]Gideon, F. C., Lt. Col. USAF (Chief, Aircraft Div., DCS/Plans and Programs), Memorandum for the Record, *AFSC Council Minutes—New Fighter Aircraft—The Advanced Tactical Fighter Program,* Nov. 5, 1981.

[39]Robinson, C. A., "USAF Reviews Progress of New Fighter," *Aviation Week and Space Technology,* Nov. 28, 1983, pp. 44–46, 51.

Chapter 6

ENGINE DEVELOPMENT

The development of the ATF engines occurred in parallel with the airframes. These engines, the Pratt and Whitney YF119 and the General Electric YF120, were developed under a series of programs, leading from design studies and technology demonstrations in the early 1980s, to prototype ground and flight demonstrators, and, finally, to proposed production configurations by 1991. That year, the Air Force chose the Pratt and Whitney F119 to power the F-22 for the 21st century.

INITIAL ENGINE STUDIES

The Advanced Tactical Engine (ATE) program officially began in May 1981 when the ATF request for information (RFI) was released to industry. All major domestic engine companies (including General Electric, Pratt and Whitney, Detroit Diesel Allison, Garrett and Williams) were invited to attend the ATF government technical briefings by the airframe companies. Because only Pratt and Whitney and General Electric continued into the Dem/Val phase, the contracted efforts by the other companies will not be discussed here.

Advances in technologies in the mid- and late 1970s, under programs such as Advanced Turbine Engine Gas Generator (ATEGG), Aircraft Propulsion System Integration (APSI), and Joint Technology Demonstrator Engine (JTDE), provided the technologies that were the basis for the contractor study engines. Contractor data submitted in the late 1970s proposed a number of innovative concepts to future propulsion needs. Pratt and Whitney's Multiple Application Core Engine (MACE) was a common core with a fan tailored for various supersonic and subsonic aircraft, including vertical/short takeoff and landing (V/STOL), ATF concepts, and even an advanced transport.

While the ATF was being defined, studies were conducted in parallel to evaluate possible propulsion systems for future Air Force and Navy aircraft. These studies, described in the following section, were essential to defining the technologies as well as the size and cycle of the ATF engines.

The advanced technology engine studies (ATES) was a Navy-sponsored program that ran from 1980 to 1982 and was not tied to a specific weapon system. The purpose was to "provide a coordinated Government/Industry long range propulsion plan which is geared to reduce the life cycle cost (LCC) of future weapon systems while at the same time providing the performance necessary to meeting advanced weapon system requirements."[1] Possible future Navy and Air Force aircraft included advanced fighters, V/STOL, and transport aircraft. Boeing, Grumman, McDonnell Douglas, and Vought were subcontractors to Pratt and Whitney to properly assess engine/airframe interactions and cycle optimization including installation effects. Boeing, Grumman, and McDonnell Douglas also were subcontractors to General Electric, with the addition of Sikorsky, which was subcontracted for marinized rotorcraft studies. The Navy was particularly interested in its VFMX (advanced carrier-based multirole fighter) aircraft, which was seen as an F-14 replacement and later evolved into the NATF.

Additional study objectives were to define the engine design parameters (engine size, cycle, cost, reliability, maintainability, and supportability) over a wide range of potential weapon system requirements; define high payoff/critical advanced technologies and the maturation methodologies; define the most cost-effective development and qualification strategy; and compare common core (such as MACE) and unique engine concepts, identifying the most cost-effective approach.[1]

In parallel to the ATES, the Air Force issued its ATF Engine RFI in June 1981 in conjunction with the ATF RFI. The following month, the Air Force assigned Robert J. May* as the program manager of the engine development effort in ASD's Aero Propulsion Laboratory (APL, now the Air Force Research Laboratory). At the time, May was a GS-13 engineer, previously assigned to the APSI and ATEGG programs. Gary A. Plourde was the Pratt and Whitney program manager (1981–1989), followed by Walter Bylciw. Dean Leonard (1983–1986) followed by Michael Brazier (1986–1991) were the General Electric program managers.

The ATFE RFI stated the desired attributes: supersonic persistence without afterburner (supercruise), STOL distances of 1500 ft, stealth, a reduced cost of ownership, and a targeted system initial operational capability (IOC) of 1993. The aircraft was anticipated to have a gross

---

* May is now the Propulsion Product Group manager in charge of all fielded U.S. Air Force engines, based at the Air Logistics Center, San Antonio, Texas.

takeoff weight of 40,000–60,000 lb, carry conformal stores, and "satisfy a twin engine air-to-air role with the off-design capability of performing a deep interdiction air-to-ground role."[2]

Responses from industry came back in early 1982. The initial studies under ATES stated that, although current and derivative engines could be developed to meet the anticipated mission requirements, it would only be with significant life cycle cost ($4.2 billion according to Pratt and Whitney analyses) and gross takeoff weight (30–50%) penalties. Thrust reversing provided significant reductions in landing distance particularly in wet or icy conditions; in fact, with thrust reversers, the landing length required was estimated to be almost unaffected by the surface condition. Stealth features were seen as achievable with a high system survivability payoff, but required additional funding for transition to a propulsion system. A joint Air Force ATF and Navy VFMX engine development program was seen as "feasible and mutually beneficial" both technically and programmatically, with a common engine core as the product. The optimum VFMX engine, however, was nominally seen as having 5000 lb more thrust and a bypass ratio (BPR) in the neighborhood of 0.8 (Ref. 2).

Supercruise was determined to be possible with

> significantly more intermediate power thrust to minimize SFC [specific fuel consumption]...achieved through proper cycle selection and power management schedules. An advanced turbojet whose high temperature capability is devoted to the supersonic regime and whose nozzle area scheduling is uniquely defined can provide a 70–100% increase in supersonic T/W [thrust-to-weight] over current engines.[2]

Increased combustor exit temperatures for higher thrust and advanced materials for lower weight were the key technologies required for this increased T/W.

The Pratt and Whitney ATES design for the ATF aircraft was a very low bypass ratio (BPR of only 0.15) counter-rotating turbofan based on work done on its ATEGG 685 and JTDE 690 demonstrators. In fact, the original internal designator was the STJ 562 ("study turbojet") indicating that it was basically a "leaky turbojet" (i.e., just enough bypass flow to provide cooling for the aft components, as opposed to a true turbofan in which the bypass air contributes substantially to the total thrust). With a thrust of about 22,000 lb, overall pressure ratio (OPR) of 24, and combustor exit temperature (CET) of 2900°F, it was to be designed for 12,300 total accumulated cycles (TACs). Compared to current generation engines (F100/F110), the acquisition cost was to be reduced to 70% and maintenance cost to 45% due to fewer stages and fewer parts.[2,3]

General Electric's ATES studies centered around a series of fixed and variable cycle engines designated as the GE16 family. They typically employed a five-stage compressor and a two- or three-stage fan configuration. A range of turbine temperatures were evaluated with single-stage high-pressure turbines and one- or two-stage low-pressure turbines. Study engine cycles reflected conventional turbine configurations (i.e., with vanes between the stages).[4,5]

PROPULSION ASSESSMENT FOR TACTICAL SYSTEMS

The propulsion assessment for tactical systems (PATS) studies ran from September 1982 to September 1983. The objectives of the studies were to "assess the propulsion needs of future U.S. Air Force and Navy fighter aircraft and to evaluate the merits of new advanced technology engines as compared to existing engines."[6] PATS also evaluated LCC tradeoffs of critical engine technologies and optimized the engine configurations for specific mission requirements.

The PATS program picked up where ATES had left off, but was augmented by the release of the initial draft Tactical Air Forces Statement of Need (TAFSON) in early 1982, which began defining the overall requirements for the ATF. The ongoing definition of ATF requirements during this period reduced (slightly) the uncertainty in engine size and allowed further definition of the candidate engines.[6]

As with ATES, selected aircraft companies were subcontractors to the engine companies for this effort. The airframe manufacturers were responsible for trade studies on the mission profile and engine integration. Because each subcontractor was allowed to choose the technologies, aircraft planforms, weapons, and, to a certain degree, performance parameters, the engine requirements for each system varied considerably. Eventually, however, a convergence was achieved to essentially one Air Force and one Navy engine size and cycle, rather than an optimized solution for each one, because aircraft size was found to be relatively insensitive to small perturbations about the optimum engine cycle. Thus, an engine cycle that was fairly close to the optimum cycles of all the different aircraft designs would not impose a great penalty on any of them.[6]

PATS also defined initial Air Force and Navy missions and allowed Pratt and Whitney and General Electric to define "optimum" new and derivative engines for these missions and refine the duty cycles. Compared to the Navy engine, the resulting Air Force engine was lower bypass and greater sea level thrust size, although the technology levels were similar. Sensitivity analyses were conducted, and an optimum Air Force/Navy cycle was developed with some penalties for each system.

Derivative engines were again studied, utilizing improved F100/F110 cores with the advanced technology fan, low-pressure turbine, and exhaust system that would have to be developed for the new engine, but they caused large penalties to weapon system size, cost, and performance, particularly the Navy system performance. With both derivative engines, there was also substantial increase in aircraft size and an inability to meet some of the maneuverability requirements. In addition, the life cycle costs were higher for these engines than for new engines based on the available technologies.

Pratt and Whitney continued refining their study engine to the revised performance needs. General Electric concluded from their studies that the variable cycle concept better matched the wide range of ATF solutions that were still being considered. Their concept provided high nonaugmented specific thrust to Mach numbers above two and flexibility to reduce SFC at subsonic cruise. The components and architecture of the variable cycle engine (VCE) had been demonstrated in a building block approach through Air Force-, Navy-, and NASA-sponsored programs in a modified variable cycle YJ101 engine core with split fan stages and in the GE23 JTDE with a core-driven fan stage (CDFS). In addition, a GE23 JTDE test was conducted in 1984, which proved the concept of a counter-rotating vaneless high-pressure turbine (HPT)/low-pressure turbine (LPT) interface.[4,5]

GROUND DEMONSTRATOR ENGINES

The name of the ATF engine program was changed to the Joint Advanced Fighter Engine (JAFE) in 1982 in recognition of a memorandum of understanding (MOU) with the Navy. In May 1983, in conjunction with the formal RFP for the concept definition phase of the ATF program, the JAFE RFP was released to Allison, General Electric, and Pratt and Whitney. Because of the security restrictions of the program, only domestic sources were sought, and these three companies were the only U.S. companies that were seen as capable of developing the large thrust class engine required. Allison did not submit a proposal, having had technical problems with their advanced development demonstrators and knowing that only two contracts would be awarded.[7,8]

In September 1983, General Electric and Pratt and Whitney were each awarded $202 million [the total amount for engine development during this period was $470 million, with the remainder going toward other government costs, including Arnold Engineering Development Center (AEDC) altitude testing and support]. The objectives were to "develop the critical advanced propulsion system technologies and support concepts" needed for the ATF propulsion system and to demonstrate those technologies and concepts on ground demonstrator engines. The engines were expected to be internally representative of possible production designs, but not flight weight or flight cleared. In an innovative approach, a fuel consumption requirement was included in the contracts.[7]

The firm fixed-price contracts were intended to run for 50 months for a combined concept definition and Dem/Val phase. The engine concepts would be evaluated by a comprehensive altitude performance and operability test and a 250-h accelerated mission test. It was planned that only one contractor would be selected for a 72-month follow-on development program to meet the aircraft schedule requirements, although the possibility of a leader–follower competition was left open. Because the development of a second engine through full-scale development (FSD) was estimated at requiring another $1.5 billion (in FY1983 dollars), it was not seen as a fiscal possibility. The two selected airframe designers were to evaluate each engine in their aircraft, but the final engine selection was to be made by the Air Force. In conjunction with the move to this demonstration phase, the engine management was changed from solely APL to a "dual-hatted" organization under May who reported to both the New Engine SPO (ASD/YZ) and APL.[7,9]

As mentioned in Chapter 3, seven airframe manufacturers were also awarded concept development investigation (CDI) contracts in September 1983: Boeing, General Dynamics, Grumman, Lockheed, McDonnell Douglas, Northrop, and Rockwell. The CDI contracts represented an important step in narrowing down the ATF concept. The earlier RFI effort had been extremely broad, with concepts falling into four major categories: extremely lightweight "numbers fighters," subsonic low observable aircraft, supersonic cruise and maneuver fighters, or high-fast fliers reminiscent of the Lockheed YF-12. The CDI effort focused specifically on an air superiority fighter with supersonic cruise and maneuver capability, reduced observables, and greatly improved reliability, maintainability, and supportability relative to existing fighters. This allowed the JAFE propulsion contractors to concentrate on engines in the 30,000-lb thrust class with technologies that would provide a supercruise capability, reduced parts count/improved reliability, and two-dimensional nozzles both for signature reduction and for a vectoring capability for improved maneuverability.

During 1984, the ATF CDI contract efforts were completed, the final TAFSON 304-83 released, and the ATF program prepared for Milestone I, which would mark the beginning of the demonstration and validation (Dem/Val) phase. The desired characteristics for the ATF at this time included: a combat radius of 500 miles (mixed subsonic/supersonic mission) or 800 miles (purely subsonic), supersonic cruise (supercruise) at Mach 1.4 to 1.5, a 2000-ft takeoff and landing distance, a gross takeoff weight of 50,000 lb, a unit cost (in 1985 dollars) of $40–45 million, and an LCC as good as or better than the F-15.

Design work on the engines proceeded, and detailed design reviews (DDRs) of the ground demonstrator engines were conducted in February and March 1985 by the program office with each JAFE contractor. The purpose of these in-depth, on-site reviews was "to determine the potential of [each] design to meet specification requirements.... The DDRs are part of the continuing approval process prior to releasing the design for engine hardware fabrication."[10] Both engine companies began fabrication of their ground demonstrator, or "XF," engines soon after these reviews.

About this time, May left ASD, and the engine management was restructured with a military officer as the program manager and a civilian as the chief engineer. The program management responsibility was officially transferred to ASD's New Engine SPO (ASD/YZ) in recognition of the fact that it was becoming a "real program." In 1987, it was moved to the ATF SPO to more closely couple the two development programs.

XF119

Taking the conceptual designs of its ATES and PATS study engines, Pratt and Whitney began the detailed design of its PW5000 (which was later designated F119 by the Air Force) once the RFP was officially released in May 1983. Advances in technologies in the late 1970s under ATEGG and JTDE had proven the capability to do more work with fewer stages (i.e., achieve a higher compression ratio per stage). Whereas the F100 engine had a 10-stage compressor, the PW5000 now achieved a higher pressure ratio with only six (see Fig. 79) as a result of improved blade aerodynamics and new materials and manufacturing processes. A single high-pressure turbine stage drove the compressor, and a single counter-rotating low-pressure turbine stage drove the three-stage fan. This design not only reduced weight and length, but it also decreased the manufacturing cost of the engine. The compressor and the second and third stages of the fan used integrally bladed rotors

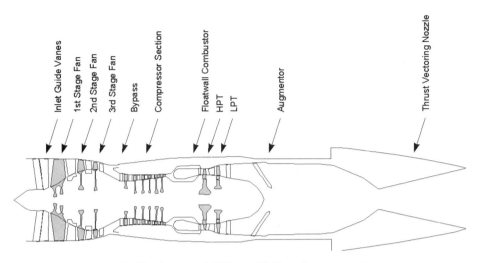

Fig. 79   Pratt and Whitney F119 engine concept.

to further reduce weight and cost and improve maintainability and performance.[11,12]

The use of a Floatwall™ combustor design allowed higher temperature and performance; Floatwall eliminates welds in the combustor, which keeps cracks from forming as a result of thermal cycling. The single-stage turbines used advanced film cooling to improve performance and durability with lower fuel consumption. The dual redundant full-authority digital electronic control (FADEC) was designed to improve safety and reliability.[13]

Fabrication of the PW5000 parts began in September 1985. Assembly began in September 1986, and the first XF119 engine, FX601, ran for the first time the next month (see Fig. 80). The second XF119 engine, FX602

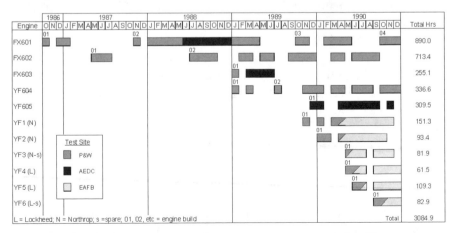

Fig. 80   The XF119/YF119 development test schedule.[18]

Fig. 81   FX602 on test in afterburner with slave nozzle.

(Fig. 81), began running in May 1987. Over the next four years, these two engines were to be the workhorses on the F119 development program, accruing over 1600 hours of run time. Engine FX601 was also tested at simulated altitude conditions at Arnold Engineering Development Center (AEDC) during the second half of 1988 for 125.5 h.[18]

The engines were tested first with an axisymmetric slave nozzle and then with a generic two-dimensional convergent-divergent (2DCD) vectoring nozzle, which was being developed to improve survivability as well as maneuverability of the aircraft. Pratt and Whitney tested FX602 in early 1988 with a 2DCD nozzle that had moveable top and bottom panels that could vector the engine thrust ± 20 deg in pitch (Fig. 82). The nozzle flaps could also close down, allowing cascade vanes on the top and bottom to vector the thrust forward for in-flight braking or for a reduced landing roll.[12,13]

XF120

The initial definition of the General Electric GE37, which became the XF120, was focused on performance for the supercruise requirement. The design was a counter-rotating variable cycle engine with a vaneless HPT-LPT interface designed at the minimum bypass ratio.[4,5,14]

Fig. 82   FX602 on test at Pratt and Whitney, West Palm Beach, Florida, in afterburner with the thrust vectoring/reversing nozzle.

A variable cycle was attained by controlling the variation of bypass ratio using a double bypass concept (see Fig. 83) in concert with the core-driven fan stage. The engine operated in double bypass at low power to gain the specific fuel consumption (SFC) benefit of higher BPR; at high power it operated in single bypass (low BPR) to achieve high specific thrust. Fan-to-core pressure matching was achieved through a variable area bypass injector (VABI) just ahead of the aug-mentor. The VABI supplied required exhaust liner cooling air and in-jected the remainder of bypass air back into the exhaust upstream of the throat area to maximize thrust potential. The engine was controlled by a three-channel FADEC.[5,14]

The two-stage fan was of blisk construction ("bladed disk," equiva-lent to Pratt and Whitney's integrally bladed rotor design). The blades were low aspect ratio to meet performance requirements, giving low leading-edge stress and high tolerance for foreign object damage (FOD) and repair blending. The five-stage blisk compressor was begun by the core-driven fan stage. This "fan stage" was followed by the sin-gle bypass flow opening.[5,14]

The double-dome annular combustor allowed for a short, efficient combustor section. The single high-pressure turbine stage was a high

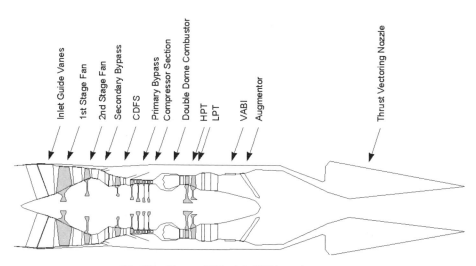

Fig. 83   General Electric F120 engine concept.

reaction design with exit swirl to match the single low-pressure turbine inlet needs. There were no stationary vanes between the turbine stages.[5,14]

The two XF120 demonstrator engines accrued 400 total test hours (see Fig. 84) in addition to component testing, including 8000 h for controls and accessories (C&A), 5500 h of structural testing, as well as 500 h of low-observables technology testing. The first, highly instrumented engine (888-010) was unaugmented and used a slave nozzle. It was tested at General Electric as well as in the altitude cells at AEDC. It first ran in May 1987, gathering 130 h at sea level and 60 h at altitude.[4,14]

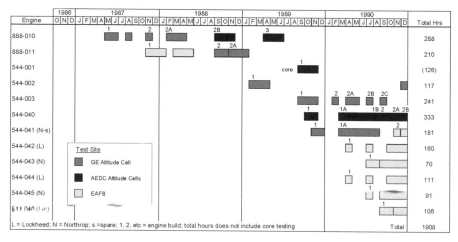

Fig. 84   The XF120/YF120 development test schedule.[4]

The second engine (888-011), tested at a General Electric site on Edwards Air Force Base (Fig. 85) because of the noise and security restrictions, demonstrated augmentor operation as well as a thrust vectoring 2DCD nozzle. It first ran in November 1987 and accrued 210 h of operation, but the integral thrust reversal system (which was similar in concept to the Pratt and Whitney design) was never tested because this requirement was deleted from the ATF program just prior to General Electric demonstrating their capability on an engine test, as explained in the next section.[4]

The XF120 testing validated the engine design concepts. General Electric had originally planned on using a carbon-carbon nozzle. However, carbon-carbon technology being explored in component testing did not mature as rapidly as expected, and the evolving ATF requirements were changed to place greater emphasis on low observables, *particularly in the aft sector,* in the RFP that was released in November 1985. General Electric responded to these developments with metal exhaust system and a larger fan for increased cooling air.[4,5]

FULL AUTHORITY DIGITAL ELECTRONIC CONTROLS

As jet engines have evolved, so have their control methods. Purely hydromechanically linked control systems began to give way in the early 1980s with the introduction of electronic engine controls (EECs)

Fig. 85   888-011 on test at Edwards Air Force Base, California.

on Pratt and Whitney and General Electric fighter engines. These control systems acted only in a supervisory function. Starting, acceleration, deceleration, and shutdown, as well as engine speed, variable vane angles, and engine bleed systems were all controlled hydromechanically. The EEC more precisely regulated the engine thrust based on pilot input, regardless of flight or ambient conditions, while limiting the engine to safe operating speeds and temperatures. The EEC had a hydromechanical backup in case of control problems or if the pilot selected hydromechanical control.

The ATF engines required a new level of control capability in order to meet their objectives. A full-authority digital electronic control (FADEC), as used on the ATF engines, was an EEC with no hydromechanical backup. It controlled all engine functions, eliminating the need for a hydromechanical backup. It reduced pilot workload during critical operations because of the control logic, which simplified power settings for all engine operating conditions. It provided consistent starting schedules, fast response to pilot requests for power, repeatable thrust settings, enhanced reliability, operability, and diagnostic capability.

The advantages of a FADEC are the following: no engine trimming is required, engine speed remains constant as the engine steady-state operating line changes because of environmental conditions and bleed demands, the controls can implement complex bleed schedules that permit the bleeds to be closed over a greater range of aircraft operation, more precise control of variable geometry vanes, more repeatable transient performance, improved starts due to accommodations of the ambient conditions, and automatic limiting of critical engine pressures and speeds.

The ATF FADECs were fuel-cooled, engine-mounted assemblies with fully redundant dual electronic channels, each with a separate processor, power supply, memory, input sensors, and output actuators. No single electronic malfunction was able to cause an engine operational problem. Each control channel incorporated fault identification, isolation, and accommodation logic.

FLIGHT DEMONSTRATOR ENGINES

As discussed in Chapters 3 and 4, the formal ATF RFP for the Dem/Val phase was officially issued in September 1985. The Dem/Val program was designed to include full- and subscale models for wind-tunnel testing and RCS measurements, and engine, avionics, and other subsystem development. These system tests, along with computer modeling and simulation, were intended to significantly reduce the cost vs a full-scale flying prototype program. Two months later, the Air Force updated the RFP with more stringent stealth goals for the ATF designs, which severely influenced the engine designs, as already noted.[9]

As a result of the February 1986 Packard Commission recommendations, the Air Force decided that each airframe contractor would now fly two prototype aircraft and would flight-test JAFE-derived engines from both engine contractors. This prototyping approach required corresponding modifications to the engine program: "The competing engine contractors will provide weapon system integration efforts, increase their special technology efforts [LO] and reorganize their AMT [accelerated mission testing] test approach to provide flight clearance engines...."[15] Initial changes in the engine program were covered in a mutually agreed-to contract modification of June 11, 1986. These changes redirected the program to ensure that flight test engines and associated certification testing were properly planned for and provided an additional $30 million to both companies to order long lead items for the additional engines.

Concurrently, the engine program name was again officially changed, this time from JAFE to the Advanced Tactical Fighter Engine (ATFE) Program, when the Navy temporarily withdrew from the engine program after the projected operational date of its F-14 replacement slipped past the year 2000. The Navy continued to participate in the ATF program through EMD source selection in early 1991 but withdrew entirely later that year.[7,8,16]

Full funding of the engine companies' participation in the ATF flight-test effort was accomplished through contract modifications signed during 1987. The modifications added $341.9 million, bringing the total value of the ATF engine contracts to this point to about $575 million each. The "ATFE Prototype Restructure" contract modifications provided for six flight-test engines from each contractor and extended the effort through December 1990. In addition, funding for the initiation of the formal FSD preliminary design effort by the two engine companies was in place by February 1988.[17]

The change to a flight-test program had a significant impact on the ATF engine development program. Flying prototype, or "YF," engine testing had only been planned to occur in EMD after engine and airframe contractor downselect. Initiation of flight-test in the Dem/Val program required the engine contractors to evaluate their configurations to decide which technologies and features could be implemented in a flight weight design and flight cleared in time for the planned flight-test program. How much advanced technology to include in the flight-test engines was left up to each contractor. Only component testing had been done up to this time. Pratt and Whitney and General Electric had to decide by the beginning of 1987 (by which time Pratt and Whitney had only 71.7 h of XF engine run time and General Electric had none) which technologies to flight-test and which to hold back for their ground demonstrator engines. Technologies that were proven in the ground tests would be eligible to be included in the FSD engine proposals.[4,18] According to Col. Piccirillo, "It's up to them how aggressive they want to be, and how much risk they want to take."

In late 1987, information from the aircraft companies indicated a heavier and higher drag aircraft than had previously been expected. The takeoff weight goal of 50,000 lb had always been very ambitious, but it was hoped that a combination of advanced technologies and careful requirements trades would allow the goal to be achieved. By 1987–1988, it was finally beginning to be accepted that the 50,000-lb goal would not be met. The combination of characteristics sought in the ATF—including large combat radius on internal fuel, supersonic cruise speed, very high maneuverability, internal weapons payload, and very low observables—simply required a larger airplane. As a result, the airframes required more thrust to meet performance requirements than was originally planned. Maximum thrust required in afterburner was increased by nearly 20% to the 35,000-lb "thrust class"; thrust level without augmentor was about 23,500 lb.[4,12]

General Electric responded to this and the change to a metal air-cooled nozzle by incorporating a 12% larger fan in its demonstrator program; Pratt and Whitney ground demonstrated a 15% larger fan to make the additional thrust, but retained the smaller fan version for its flight test design.[4,5,13]

NOZZLES

In December 1987, thrust reversing was dropped. After a great deal of study, the prime contractors concluded that the operational benefits would not justify the added weight, cost, and maintenance required. Nozzle cooling inadequacies were discovered during detailed design and ground engine testing for the F-15 STOL and Maneuver Technology Demonstrator (S/MTD). The cooling requirement during thrust reversing greatly increased the weight of the nozzles. These findings caused the Air Force to relax the ATF landing requirement from 2000 to 3000 ft, allowing thrust reversing to be deleted. Development of thrust vectoring technology was continued, however, as it was expected to enhance the ATF maneuverability at low speeds and again above about Mach 1.4. In the middle of the flight envelope, the aerodynamic controls are more effective, but at higher speeds changes in the airflow reduce their control power, and vectoring improves turning rate by about one third.[7,12,19]

The two-dimensional nozzles (seen in Figs. 86 and 87) were tailored to each competing airframe design to optimize the performance and survivability for that airframe, but the nozzles were common from the turbine aft face up to the throat (station A8, the point of minimum cross-sectional area). The engines were developed as government furnished equipment (GFE), but the weapon systems contractor (WSC) unique divergent sections (from stations A8 to A9) were developed as contractor furnished equipment (CFE). Lockheed's ATF design concept included thrust vectoring nozzles with moveable top and bottom flaps. Northrop's design had fixed bottom surfaces and moveable top flaps for area control only. The nozzle could not vector, being constrained by the exhaust trough design of the YF-23.[4]

YF119

Only slight changes were made from the XF119 to the original YF119 design (Fig. 88). Pratt and Whitney switched to the burn resistant titanium alloy C in the nozzle and augmentor to reduce weight and improve durability and safety (more fire resistant in that it does not self-combust in the F119 flight envelope). Hard smooth abradables and coated blade tips were also added to improve performance and durability. Alloy C was also used for the initial static stages of the compressor section. The turbine rotors were fabricated from single-crystal materials and dual heat treated with properties that varied along the radius of the disk. This allowed rim areas to have a coarse grain for improved damage resistance and a bore with a finer grain that yielded increased strength and cycle fatigue resistance.[13,20]

When Pratt and Whitney realized they needed to provide additional thrust for the ATF aircraft, they elected not to try to implement the design in time for the flight tests. Instead, it was to be verified during ground tests to be completed prior to proposal submission. Inserting this design into the flight-test engines could have spelled disaster if it had displayed operability, reliability, or maintainability problems.[13,21]

The fan diameter, therefore, was slightly increased, which provided a 15% increase in fan air flow and an increase in bypass ratio to 0.30 (from 0.25, see Table 7). Although some aerodynamics and blades were also changed, the rest of the engine, including flow paths, spacings, and rotor sizes remained unchanged. The new fan design was extensively rig-tested at Pratt and Whitney and the Wright Laboratories Compres-

Fig. 86   Pratt and Whitney's Lockheed (left) and Northrop (right) nozzle frames. They had common round-to-square transition ducts (top), which are also seen at the bottom of the nozzle outer walls.

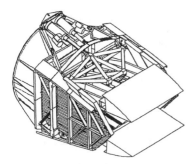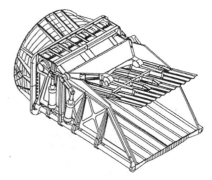

Fig. 87   General Electric's designs for Lockheed (left) and Northrop (right) nozzles. An additional piece (not shown) provided a chevron trailing edge to the top and bottom to match the aircraft angles.

Fig. 88   YF119 with YF-22 nozzle demonstrating thrust vectoring at Edwards Air Force Base, California.

Table 7   YF119 configuration

Fan stages	3
Compressor stages	6
High-pressure turbine stages	1
Low-pressure turbine stages	1
Bypass ratio	0.25
Thrust class (in afterburner)	30,000 lb

sor Research Facility (CRF) at Wright–Patterson Air Force Base, Ohio, in early 1990. The full-up refanned XF119 engine (FX601-04) was tested at West Palm Beach during the final months before the proposal was submitted. The thrust increase provided by the new fan exceeded the required increase by 20%.[13,21]

The first flight-prototype engine was tested on January 1989. In October, the first Northrop nozzle was tested. The following April, the two flight engines, YF1 and YF2, were delivered to Northrop, with a spare delivered the following month. Also during April, the first Lockheed nozzle began testing, with engine deliveries (YF4 and YF5) in July and August 1990, and the spare that October. Also in August, engine YF604-2 completed its accelerated mission testing (AMT), and engine YF605-1 completed flight clearance testing at AEDC. Over 3000 h were accrued in the XF119/YF119 engine test program.[13,18]

YF120

General Electric initiated its YF120 engine (Fig. 89) design in February 1987 and achieved first engine to test (FETT) in February 1989 (24 months). The design process continued to evaluate variations in design to address airframe contractor mission needs and competitive pressures. This resulted in two compressor configurations being engine tested; these two designs were designated Block I and Block II. The final result was that the YF120 engine configuration flight-tested was very close to the final proposed EMD design, including the all-blisk compression system. Proposed F120 production engine thrust levels were demonstrated and flown.[4,5]

Assets used in the YF120 test program consisted of the XF120 engine 010 used to explore core and fan operability, a YF120 engine core tested at altitude at AEDC (001), two factory development engines (002 and 003), two flight clearance engines (040 and 041), and five flight-test engines (042 to 046). Engine 041 was used for endurance test-

Fig. 89   YF120 with YF-22 nozzle.

ing, and then refurbished as a flight spare that could be configured for the YF-22 or the YF-23 airframes. The first General Electric/Northrop nozzle was tested in December 1989 on engine 041, and the first Lockheed nozzle began testing with engine 040 in May 1990. The engines for each airframe were common through the nozzle throat, but each engine had a unique divergent section and control logic. The YF120 accrued 1089 engine factory test hours, including 915 h at simulated flight conditions in General Electric and AEDC altitude cells.[4]

The YF120 design (see Table 8) employed a fan and airflow size 12% larger than that of the XF120. This size increase was in response to the need for more cooling air for the metal exhaust system and the increased ATF thrust requirements. Overall pressure ratio was 22. Durability was addressed using thermal barrier coatings (TBCs) and tailored cooling air distribution. AMT endurance testing was conducted at a turbine temperature that allowed a margin for safety and engine-to-engine variations. As a result of the short, compact hot section with the vaneless high-pressure/low-pressure turbine concept, there was still 30% less hot section cooled surface area than the General Electric F110 engine. Variable cycle engine (VCE) features were simplified from the XF120 design, and maintainability requirements were rigidly enforced during the design process.[4,5]

Table 8   YF120 configuration

Fan stages	2
Compressor stages	5
High-pressure turbine stages	1
Low-pressure turbine stages	1
Bypass ratio	0.32
Thrust class (in afterburner)	30,000 lb

The F120 EMD configuration was tested for 21 h in December 1990, using the Block I engine 544-002 with hollow first-stage fan blades, a passive fan bypass door, and Ada software. The F120 EMD design incorporated component improvements that allowed the engine to operate at the required performance levels with lower turbine temperatures than required to meet YF120 flight-test thrust levels. As with the F119, the F120 EMD design was changed to include either fuel or shared airframe/engine hydraulics. Over 1900 h were accrued in the XF120/YF120 engine test program (see Fig. 90).[4,22]

ENGINE FLIGHT-TESTING

The YF-23 made its first flight on August 27, 1990, powered by the Pratt and Whitney YF119, and the YF-22 followed on September 29, 1990, powered by the General Electric YF120. The F119 powered the YF-22 on 31 sorties for 153.6 engine flight hours (Fig. 91) and propelled the YF-23 for 34 flights and 81.4 engine flight hours. The General Electric–powered YF-22 made 43 flights accruing 105.6 engine flight hours. The YF-23 with the YF120 made 16 flights with 42 engine flight hours over seven weeks.[9]

Flight-testing validated the expectations of supercruise. According to published reports, a YF120 powered YF-22 flew in formation with an F-15 at Mach 1.58 at 40,000 ft and used 30% less fuel than the non-supercruising F-15. The small-fan YF119 flight demonstrator was only capable of pushing either airframe in supercruise to about Mach 1.43. The YF120 also achieved the maximum speeds, reportedly over Mach 2 in level flight for both the YF-22 and YF-23 airframes in afterburner.[19]

These extremely fast-paced test programs were made possible by advances in the collection, transmission, and processing of flight-test data and also by minimizing government control and letting the contractors run their flight-test programs, another outgrowth of the Packard Commission. Traditionally, flight-testing two new airframes with two new engine designs would have been fraught with delays. But, by having a proven flight clearance program for the engines and a greater degree of contractor autonomy, the ATF program had one of the most ambitious and successful flight-test programs ever.[9]

Instead of trying to meet detailed government-specified flight points, the contractors were able to gather the data they felt was necessary for their proposals. These data were used to validate the critical areas that each contractor felt essential to validate its aircraft's predicted performance. This meant that the flight envelope could be cleared quickly by clearing a speed-altitude path from low-speed/low-altitude to high-

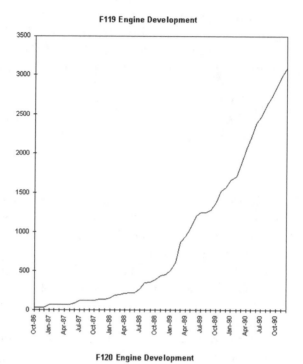

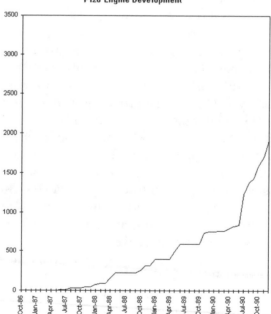

Fig. 90   ATF engine development test hours.

Fig. 91 YF119 engine run-up testing on the YF-22 at Edwards Air Force Base, California.

speed/high-altitude to the area of interest: supercruise. The primary government-imposed requirement was that each aircraft fly at least once with each company's engine.

The Air Force had safety authority, but the contractors had technical cognizance. Decisions could thus be made more rapidly as a result of the highly focused contractor management and their inherent knowledge of their aircraft and company strengths and weaknesses. Northrop and Lockheed also contributed nearly 50% of the cost of the aircraft development. With such a large investment, they had a greater vested interest and greater say over the flight-test program.

ENGINE SUPPORTABILITY FEATURES

The ATF engines were designed with supportability in mind. Both designs had 40% fewer parts than their predecessors (the F100 and F110), fail-operational FADEC, and refurbishment interval of 4325 TACs for the hot sections and twice that for the cold sections (which was anticipated to be about 15 years of service life). Pratt and Whitney made maintainability one of their primary areas of emphasis. Line replaceable units could be removed in less than 20 min with 11 standard hand tools and were only "one deep." The engines were designed to accommodated maintainers from the 5th percentile female to the 95th percentile male. On-aircraft maintenance could be conducted while wearing hazardous environment protective clothing.

Built-in test and diagnostics were integrated with the aircraft support system, eliminating the need for a special engine support system. Lock-wire was eliminated, and torque wrenches were no longer required for "B" nut installations. The engine was designed with built-in threadless borescope ports, axially split cases, oil sight gauges, and integrated diagnostics. Other improvements were a modular design (see Fig. 92), color-coded harnesses, interchangeable components, quick disconnects, automated integrated maintenance system, no component rigging, no trim required, computer-based training, electronic technical orders, and foreign object damage and corrosion resistant. These advances were intended to reduce operational level and intermediate level maintenance items by 75% and depot level tools by 60%, with a 40% reduction in average tool weight.[13]

ENGINEERING AND MANUFACTURING DEVELOPMENT ENGINE SELECTION

The long lead protection package for the ATFE engineering and manufacturing development (EMD) phase was begun in March 1990, allowing preliminary design and material release. A contract of $290 million was issued to each engine company. EMD proposals from the two ATF airframe teams were submitted on December 31, 1990, and from the two engine companies on January 2, 1991.[4,9]

The General Electric engine design demonstrated higher super-cruise and maximum speeds for both the YF-22 and YF-23 aircraft. The YF120 engine was also used for most of the high angle-of-attack (AOA) testing by both ATF contractor teams, demonstrating operability at a simulated 110-deg AOA in ground tests and 60-deg AOA at 80 kn in flight with active vectoring on the YF-22. General Electric set high performance during the flight test program as a major goal in its competitive approach.[4,5] However, the variable cycle

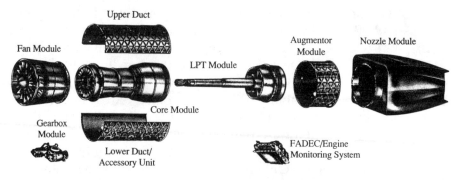

Fig. 92 The modular engine concept (F119 shown).

design increased the perceived level of risk associated with the F120 engine.[21]

Pratt and Whitney, on the other hand, attempted to demonstrate a more measured approach to introducing technical risk. In contrast to General Electric, Pratt and Whitney did not pursue a variable cycle engine, despite its experience with the variable cycle J58 for the SR-71, or a vaneless high-pressure/low-pressure turbine interface, despite the potential benefits of that feature. From the onset Pratt and Whitney wanted its flight-test engines to demonstrate a mature, well-tested easily maintained design. Consequently Pratt and Whitney accumulated over 50% more test hours than General Electric.[21,22]

On August 3, 1991, when the Lockheed F-22 team was awarded the EMD contract by ASD, Pratt and Whitney was simultaneously awarded a $1.375 billion contract. Pratt and Whitney was contracted for nine development engines, 8900 engine verification test hours, 7400 nozzle verification test hours, 33 flight-test engines, full on-site support of the flight-test program, support system development, and computer-based training system development.[13] The engine production baseline was at that time expected to be over 2000 powerplants (including spares) worth more than $12 billion to Pratt and Whitney.[13]

Pratt and Whitney proclaimed its design to be low risk using proven technology incorporated into a rugged design. It demonstrated performance margins with the larger fan (current estimate vs specification): 10% thrust (unaugmented), 140-lb weight, 2% fuel consumption, and 20–35% stall margin. It also presented a management plan and development schedule that was considered extremely responsive to customer needs.[21]

Pratt and Whitney claimed that, based on its studies, a variable cycle engine was not required for the ATF mission and added unnecessary weight and complexity. This could be translated into higher risk and greater development and/or maintenance money required. Pratt and Whitney's approach was a continued emphasis on introducing a measured, acceptable level of risk into the development program.[21,22]

ENGINE ENGINEERING AND MANUFACTURING DEVELOPMENT AND BEYOND

Development of the F119 engines continued under the F-22 EMD program, with over 8000 h of total test time on the XF/YF119 lineage of engines completed to date (June 1998). The Pratt and Whitney–powered YF-22 flew again later in 1991, but its testing was not to collect additional engine data. The flight-test YF119 engines continued to fly, and the YF-23 F119s were removed and refitted with proper logic for flight use; but after 39 additional flights (61.9 h) the aircraft was damaged in a gear-up hard landing on April 25, 1992. The engine critical design reviews (CDRs) were held in June 1992. The first F119 EMD engine began testing on December 17, 1992. The following month, the F119 engine program was rephased after funding reductions, so that Pratt and Whitney would now only build 27 (rather than 33) flightworthy EMD engines. Assembly of the first flight-test engine was completed on July 9, 1996; both EMD F119 first flight engines were on dock at Lockheed a year before the first flight in September 1997. Derivatives of the F119 engine will also power the competing Boeing X-32 and Lockheed Martin X-35 concept demonstrator aircraft for the Joint Strike Fighter (JSF) around the year 2000.

The primary differences in the F119 configuration (see Fig. 93) from the YF119 were the increased diameter low-aspect ratio fan for increased thrust and efficiency and special technologies to improve stealth and survivability. The afterburner was reduced from four zones to three to reduce weight and decrease cost and complexity. The compressor aerodynamics were improved, and the entire compression system was changed to integrally bladed rotors, whereas only second- and third-stage integrally bladed rotors had been flight-tested. Air cooling tubes on the outer nozzle walls were also eliminated in favor of internal air passages.[13,20]

The Y119 (and the YF120) employed an integral high-pressure hydraulic actuation system as an artifact of early ATF system studies, which indicated an interest in shared engine/aircraft hydraulics. The EMD design employed a more desirable fuel-powered actuation system because the requirement for shared hydraulics had been eliminated.[20]

Development of the F120 engine has also continued, although to a lesser degree. The Advanced Research Projects Agency's (ARPA) advanced short takeoff and vertical landing (ASTOVL) program used a YF120 in its test program, running for 21 h. In November 1995, the JSF

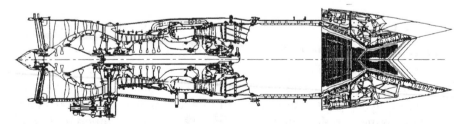

Fig. 93 F119 engine cross section (low-observable afterburner not shown).

Program Office awarded General Electric a contract for the first phase of development of an alternate engine to compete with the JSF F119 derivatives during production. The derivative of the F120 engine is now under contract to undergo a core test in 2000. A full turbofan engine test of the selected JSF concept is expected to be conducted in 2002.[4]

OBSERVATIONS ON ENGINE DEVELOPMENT

Compared to previous generations of engine development, the ATF program started out from the very beginning doing the right things to reduce technical risk prior to EMD. The original engine ground-test development would have resulted in over 500 h of run time for each contractor, half of which would have been the accelerated mission test. In contrast, the Pratt and Whitney F100 engine, developed for the F-15, was selected over the General Electric contender on the basis of two XF100 ground demonstrator engines that were run for a combined total of only 105 h before entering full-scale development. In contrast to the F100 development program, RM&S, durability, operability, and affordability were just as important in the ATFE development effort as basic performance.

There were several things that adversely affected ATF engine development plans. They can be summarized as "requirements changes." The greatest impacts were caused by the need for additional thrust, the change from a purely ground-test program to a flight-test program, and the increased requirements for aft-sector stealth. The advantages gained by early engine development can be muted by later and often significant changes in requirements. Such changes seriously affect development plans and need to be anticipated as early as possible and implemented only if truly necessary. However, the engine companies were given considerable leeway in designing and conducting their individual development efforts. This helped make the ATFE program successful despite these changes.

Engine thrust needs have to be anticipated as accurately as possible. Airframes historically have increased in weight during development, and engines are continually asked to offset these increases by providing more thrust. In the ATF program the Air Force gave the contractors an aircraft weight goal of 50,000 lb, despite earlier contractor and in-house ASD trade studies showing that this goal was highly optimistic. When it later became obvious that this goal could not be met, the engines had to be redesigned for increased thrust to meet performance requirements. If the most likely aircraft weight had been acknowledged earlier, there would have been less impact on the engine development program. It should be noted, however, that the aggressive aircraft weight and cost goals were issued in an attempt to minimize unnecessary growth; without this strong up front emphasis, the aircraft could have ended up being even larger.

Increased stealth requirements were detrimental to the original engine development plans. Again, convergence on overall system requirements needs to be accomplished as early as possible, and necessary changes must also be done as early as possible. General Electric's early plans to use a carbon-carbon nozzle were terminated when the increased rear-aspect stealth requirements were published in the revised Dem/Val RFP in November 1985. This necessitated a switch to a metal exhaust system with consequences in additional airflow needed to cool it.

Recommendations by the Packard Commission also resulted in significant changes to the ATF and JAFE/ATFE programs. The draft Commission recommendations were published prior to the beginning of the Dem/Val phase (see Table 9), allowing a minimum amount of time to change the focus of the program. With the XF engine ground demonstrators having already validated the basic engine concepts of each company, the YF engines were changed to "flight clearance" and "flight-test" engines. This change, however, mainly required the fabrication of additional assets as the YF engines were intended to be flightworthy (but not flight cleared) demonstrators from the onset. Thus, because of the intention to have a realistic ground-based competition, the change to a flight-test-based competition was not as traumatic as it might have otherwise been.

The very successful engine development and flight-test program was conducted in a highly competitive, highly independent environment. The companies were given great latitude to determine what was tested. The Air Force had safety cognizance, but not technical cognizance, leaving the contractors to set the technical goals and determine how they would achieve them. The companies gathered the data they felt were necessary for their proposals, rather than checking off a list of government requirements. With the government's concurrence, Pratt and Whitney and General Electric decided what approach to take, which technologies to test, how many engines should be tested, and how test hours should be allocated. Consequently, Pratt and Whitney built and tested fewer engines but accomplished significantly more test hours than did General Electric.

Technology programs are vital. ATEGG, JTDE, and other programs now combined under the Integrated High Performance Turbine Engine Technology (IHPTET) Program developed the technologies in

Table 9   Advanced Tactical Fighter Engine development chronology

Date	Action
June 1981	ATF RFI issued
May 1983	ATF CDI and ATFE RFPs issued
Sept. 2, 1983	ATF CDI contracts (~$1 M each)
Sept. 30, 1983	ATFE contracts awarded ($202 M)
Dec. 1984	ATF SON released
Sept. 1985	ATF Dem/Val RFP issued
Nov. 1985	RFP mod: increased stealth goals
Feb. 1986	Draft Packard Commission Report
April 1986	Dem/Val proposals submitted
May 1986	RFP mod: flying prototypes
June 1986	Engine contract mod for flight demo long lead items ($30 M each)
Oct. 1986	XF119 first engine to test (FETT)
Oct. 1986	ATF Dem/Val contracts awarded
Mar. 1987	XF120 FETT
June 1987	Engine flight demo contracts ($342 M)
Late 1987	Need for additional engine thrust acknowledged
Dec. 1987	Thrust reversal deleted
Jan. 1989	YF119 FETT
Feb. 1989	YF120 FETT
Mar. 1990	FSD engine long lead contracts ($290 M)
Aug. 1990	YF119 receives flight clearance
Aug. 27, 1990	YF-23/YF119 first flight
Sept. 18, 1990	First YF119 supercruise on YF-23
Sept. 24, 1990	YF120 receives flight clearance
Sept. 29, 1990	YF-22/YF120 first flight
Oct. 26, 1990	First YF-23/YF120 flight
Oct. 30, 1990	First YF-22/YF119 flight
Nov. 1990	FSD RFP issued
Nov. 3, 1990	First YF120 supercruise on YF-22
Nov. 15, 1990	First YF-22/YF120 thrust vectoring
Nov. 23, 1990	First YF119 supercruise on YF-22
Dec. 1, 1990	First YF-22/YF119 thrust vectoring
Dec. 10, 1990	YF-22/YF120 high-AOA tests begin
Dec. 28, 1990	YF-22/YF120 max speed (Mach 2+)
Dec. 31, 1990	WSC EMD (FSD) proposals
Jan. 2, 1991	Engine EMD (FSD) proposals
April 23, 1991	ATF/ATFE winners announced
Aug. 3, 1991	EMD contracts awarded

the 1970s that would eventually be used in the ATF engines. These programs are essential for developing new and derivative engines.

REFERENCES

[1]"Advanced Tactical Engine Studies (ATES), Final Report," Pratt and Whitney, West Palm Beach, FL, Sept. 10, 1982. (P&W proprietary; all information used has been approved for public release.)

[2]Plourde, G., "Advanced Tactical Fighter RFI Response," Pratt and Whitney, West Palm Beach, FL, Feb. 1982. (P&W proprietary; all information used has been approved for public release.)

[3]Adams, A., "Advanced Technology Engine Studies—Final Review," Pratt and Whitney, West Palm Beach, FL, Feb. 23, 1982. (P&W proprietary; all information used has been approved for public release.)

[4]"YF120 Engine Familiarization Workshop Brief," General Electric, Evendale, OH, June 12, 1996. (GE proprietary; all information used has been approved for public release.)

[5]Flynn, J., and Eismeier, M., "Advanced Tactical Fighter Engine Development Program Comments on Proposed ANSER Paper," General Electric, Evendale, OH, June 13, 1997.

[6]"Propulsion Assessment for Tactical Systems, Final Report, Vol. 1, Program Summary," General Electric, Evendale, OH, Sept. 1986. (GE proprietary; all information used has been approved for public release.)

[7]DOD Acquisition—Case Study of the Air Force Advanced Fighter Engine Program, General Accounting Office, GAO/NSIAD-86-45S-13, Aug. 25, 1986.

[8]"F119 Advanced Tactical Fighter Engine Program," Program Schedule, Pratt and Whitney, West Palm Beach, FL, Jan. 15, 1991.

[9]Abrams, R., and Miller, J., Lockheed (General Dynamics/Boeing) F-22, Aerofax Extra, No. 5, 1992.

[10]Weekly Activity Reports (WARs), from ATF System Program Office to the Commander, Aeronautical Systems Div., March 6–20, 1985.

[11]"An Interim Report to the President," President's Blue Ribbon Commission on Defense Management, Feb. 28, 1986.

[12]Sweetman, B., YF-22 and YF-23 Advanced Tactical Fighters: Stealth, Speed and Agility for Air Superiority, Motorbooks International, Osceola, WI, 1991.

[13]"Prosposal for Advanced Tactical Fighter Engine/Navy Advanced Tactical Fighter Engine, F119 Advanced Tactical Fighter Engine, Executive Summary, Vol. I," Pratt and Whitney, West Palm Beach, FL, Jan. 2, 1991. (P&W proprietary; all information used has been approved for public release.)

[14]"PPSIP Master Plan," General Electric, Evendale, OH, June 30, 1990. (GE proprietary; all information used has been approved for public release.)

[15]WARs, May 1–7, May 22–28, and June 12–18, 1986.

[16]Harris, W. R., "History of ATF/AX Engine Development Programs," NAVAIR Doc. AIR-53511B/23421, Jan. 6, 1993.

[17]WARs, Feb. 26–March 4, Oct. 22–28, and Dec. 24–30, 1987.

[18]"F119 Experimental Demo/Proto Engine Run Time History," Pratt and Whitney, West Palm Beach, FL, Jan. 1994. (P&W proprietary; all information used has been approved for public release.)

[19]"ATF Prototypes Burn One-Third Less Fuel in Supercruise," *Aviation Week and Space Technology,* Dec. 10, 1990.

[20]"F119 Configuration Reflects Balanced Design, Lessons from F100 Program," *Aviation Week and Space Technology,* Nov. 18, 1991.

[21]Kandebo, S. W., "Pratt ATF Engine Victory Could Yield 1,500 Power-plants," *Aviation Week and Space Technology,* April 29, 1991.

[22]Bond, D. F., "Risk, Cost Sway Airframe, Engine Choices for ATF," *Aviation Week and Space Technology,* April 29, 1991.

Chapter 7

Navy Participation and the Navy Advanced Tactical Fighter

At approximately the same time that the ATF program was being formulated, the Navy was pursuing a program for an Advanced Tactical Aircraft (ATA) intended primarily for the interdiction role. The ATA would replace the Navy's A-6 aircraft. Starting in the summer of 1984, the ATF program maintained an interface with representatives from the Naval Air Systems Command (NAVAIR) to exchange information on the two programs and in particular to coordinate technology development activities. Both services agreed that many or the same technologies were critical to both aircraft, particularly in the areas of low observables and avionics.[1] The coordination of technology efforts and identification of opportunities for common development were formalized in a memorandum of understanding (MOU) that was signed in April 1985 by the respective commanders of ASD and NAVAIR.[2]

In 1985–1986, Congress floated the idea of combining ATF and ATA into one program. However, neither service wanted to repeat the TFX/F-111 mistake of mixing the air superiority role with the air-to-surface role in a single aircraft.[3] The Air Force program had focused on air superiority, and the Navy program on deep strike. Nevertheless, the Air Force would eventually need to modernize its strike capability, and the Navy its air superiority capability. Congress directed that each service consider the other's program in addressing these needs. Specifically, the Senate Committee on Armed Services report on the National Defense Authorization Act for fiscal year 1987 stated: "Since the Navy must eventually replace the F-14 as well as the A-6, and the Air Force must eventually replace its F-111s along with its F-15s, the committee believes it is essential that the designs selected for the ATF and ATA [Advanced Tactical Aircraft] anticipate these additional cross-service requirements."[4]

Accordingly, in March 1986, the Secretaries of the Air Force and the Navy reached an agreement that the Air Force would evaluate the ATA (or a derivative) as a replacement for the F-111 and that the Navy would likewise evaluate the use of a derivative of the ATF as its replacement for the F-14. From 1986 until 1988, one Navy official was stationed at the Air Force's ATF Program Office to observe and monitor the Dem/Val phase of the program. A tentative operational requirement for a naval ATF variant was established in June 1987.[4] Also, in March 1987, the Navy signed an agreement with the Air Force and the Army to identify and develop common avionics for the Navy's ATA, the Air Force's ATF, and the Army's LHX (Light Helicopter Experimental), respectively. The Joint Integrated Avionics Working Group (JIAWG) was established to develop common avionics specifications and standards for these systems and their eventual derivatives.[5]

There remained congressional concerns regarding the extent to which the Navy and Air Force were actually committed to each other's ATF and ATA programs, respectively. The National Defense Authorization Act for fiscal years 1988 and 1989 made the Air Force's use of fiscal year ATF funding contingent upon the Air Force certifying that the ATF designs

> ...are capable of accepting physical and structural modifications necessary to satisfy fully the requirements of the Navy concerning catapults and arresting gear, and...that a major source selection criteria for full scale development and production will be the extent to which the contractor's proposals for the Navy-variant of the advanced tactical fighter meets fully the requirements of the Navy.

In response to this and other congressional direction, the Air Force and the Navy agreed in January 1988 that the Navy would participate in the selection of the ATF airframe and engine contractors for the EMD and production phases of the program. "The agreement provided that the Navy's requirements will be a major criteria" in the source selection, according to a GAO report on the Navy's participation in the ATF program. The report further notes that "the Navy's major issue in the selection process is the aircraft's suitability for carrier operations."[5]

To support the increased level of Navy participation, a Navy ATF program office was established in August 1988, collocated with the Air Force's ATF office at Wright–Patterson Air Force Base. The Navy's ATF office consisted of a program manager, a deputy program manager, and an assistant. Engineering support was provided by NAVAIR.

In September 1988, the Navy contracted with the ATF competitors (through modifications to the existing ATF contracts) to develop preliminary designs for Naval variants of their aircraft and to assess the carrier suitability of these variants as well as other mission capabilities that were specific to the Navy. The NATF was to be capable of operation from CV-63 (*Kitty Hawk*) and subsequent classes of aircraft carriers. In April 1989, a second set of contract modifications established specific design goals for the Navy variant, including a maximum take-off gross weight of 65,000 lb, a maximum carrier landing weight of 52,000 lb, maximum length and maximum wings-folded deck space requirement each equal to or less than an F-14. The maximum takeoff weight was derived from a goal of being able to lift two NATFs simultaneously on aircraft carrier elevators, which have a total capacity of 130,000 lb.[5]

SIMILARITIES AND DIFFERENCES

Many points in common were noted between the Air Force's ATF and the Navy's ATF requirements. Both services would require an aircraft with two engines, day/night all-weather capability, significant supersonic endurance, and reduced observability, whose primary targets would be hostile aircraft. However, differences also existed, both in the basing requirements and in the mission performance requirements. The basing considerations would impose additional requirements for the airframe structure and landing gear of the Navy variant, as well as requiring excellent low-speed handling qualities and visibility for carrier approach. There are also environmental considerations (electromagnetic environment, salt spray and other corrosives, etc.), supportability considerations, and limits on the size of carrier-based aircraft.

In mission requirements, the Air Force's requirements to maintain air superiority included supersonic cruise, very low observables, and high maneuverability. The Navy's fleet air defense mission placed more emphasis on loiter time and on the ability to detect and prosecute its targets at long range (hence the performance of the fighter's own radar is more important than its ability to avoid detection by hostile radars, for the Navy). Furthermore, the Navy required air-to-surface capability to maximize the operational flexibility provided by the limited number of aircraft in a carrier air wing.[5] (This requirement was reinforced by Gulf War experience.) Specific weapons required for the NATF included the advanced air-to-air missile (AAAM), a projected replacement for the AIM-54 Phoenix; the AGM-88 high-speed anti-radiation missile (HARM); the AGM-84 Harpoon anti-ship missile; and the advanced interdiction weapon system (AIWS) glide bomb.[6]

For these reasons, the greatest commonality was expected in engine and avionics, whereas the greatest differences would be in the airframe. The NATF would probably require a modified nose section for improved visibility and possibly a larger radar; a different wing for improved lift at low speeds (often pictured as a swing-wing in artist's conceptions) and the ability to fold for storage; a different tail for improved control at low speeds; and stronger landing gear and airframe structure. These differences were expected to result in an empty weight approximately 4000 lb heavier than the Air Force's ATF. However, the engine, avionics (with the possible exception of radar), and many subsystems would be common (although the optimum engine for the Navy mission really would have had a higher bypass ratio than the U.S. Air Force ATFE). Approximately 44% of the unit flyaway cost typically is in the engines and avionics; therefore significant savings were expected if the NATF could be successfully developed. In fact, the Navy expected to spend a total of $8.5 billion on RDT&E for the NATF, substantially less than it would cost to develop a comparable aircraft from the ground up.[5]

ACQUISITION STRATEGY

The Navy would use the same airframe and engine contractors as the Air Force. This was apparently decided in early 1988.[3] However, the Navy variants would not be prototyped during the competitive Dem/Val phase of the ATF program. Navy input to the source selection would therefore be based upon the design studies and assessments, substantiated by wind-tunnel tests of the NATF variants, which were to be accomplished prior to the source selection.[5]

For this reason, it was intended that NATF Dem/Val would then continue for approximately two and a half years after the source selection with a full-scale development decision in the fourth quarter of 1993. NATF first flight would be in January 1997 with a low-rate production decision in the fourth quarter of 1999 and full-rate production decision in the fourth quarter of 2001, approximately four years after the Air Force's ATF.[5]

NATF funding was $1.9 million in fiscal year 1988 and $65 million in fiscal year 1989 (appropriated by Congress). The first Navy request for NATF funding was for a further $65 million in fiscal year 1990, followed by $65 million in fiscal year 1991, $100 million in fiscal year 1992, $99 million in fiscal year 1993, and $395 million in fiscal year 1994 (which would be the first year of EMD). Total RDT&E funding for the NATF was projected to be $8.5 billion, and procurement costs $66.1 billion, for a production run of 618 NATFs (compared to Air Force estimates of $13.5 billion RDTE and $67.2 billion procurement for 750 Air Force ATFs).

In March 1990, it was noted that, "The Navy is uncertain if it will continue development and eventual procurement of the NATF.... The extent to which the contractors can satisfy the Navy's requirements from a derivative design of the ATF is a key factor in whether the Navy remains committed to the program."[5]

The Major Aircraft Review, launched in early 1990 under Secretary of Defense Richard Cheney, reduced the peak production rates of both the ATF and NATF, from 72 to 48 and from 48 to 36, respectively. This had the effect of substantially increasing the program cost. In August 1990, Admiral Richard Dunleavy, who was in charge of Navy aircraft requirements, stated that he did not see how the NATF could fit into any affordable plan for Navy aviation.[6] The ATF Dem/Val flight-testing was accomplished from August 1990 through January 1991, but this had little impact on the prospects for an NATF. In early 1991, consideration of the NATF was finally dropped after it was determined that the F-14 could meet the Navy's air superiority needs through 2015. There was, at least initially, an option to restart the NATF program in fiscal year 1997, although that has since been abandoned.[7]

At approximately the same time that the NATF was dropped, the Navy's A-12 program was also canceled, creating a need for a new survivable strike/interdiction aircraft. Air Force Secretary Donald Rice reportedly stated that an ATF derivative was under consideration to meet this need.[3] To study the various alternatives, the joint service Attack-Experimental (A-X) program was formulated during 1991. Five consortia were awarded contracts of $20 million each on December 30, 1991, for initial A-X studies and risk reduction. The F-22 team of Lockheed/Boeing/General Dynamics was one of the five and reportedly proposed an F-22 derivative for the A-X. In late 1992, a requirement for improved air-to-air capability was added to the A-X, leading to the new designation Attack/Fighter-Experimental (A/F-X).[8] Requirements for the A/F-X were extremely ambitious.

The 1993 bottom-up review canceled the A/F-X and initiated the Joint Advanced Strike Technology/Joint Strike Fighter program to fill the A/F-X requirement for the Air Force, Navy, and Marine Corps. This terminated for the time being any consideration of a naval ATF derivative, although Lockheed has continued to propose naval versions of the ATF, along with Suppression of Enemy Air Defenses and foreign military sales versions. The 1997 Quadrennial Defense Review (an update of the bottom-up review) stated that an F-22 derivative should be considered as a replacement interdiction aircraft taking over the role of the F-117 and F-15E in the next century.

REFERENCES

[1]Weekly Activity Reports (WARs), from ATF System Program Office to the Commander, Aeronautical Systems Div., July 4, 1984.

[2]Ferguson, P. C., *Advanced Tactical Fighter/F-22 Annotated Chronology* (Draft), Office of History, Aeronautical Systems Center, Wright–Patterson AFB, OH, Aug. 1996.

[3]Twigg, J. L., *To Fly and Fight: Norms, Institutions, and Fighter Aircraft Procurement in the United States, Russia, and Japan,* Ph.D. Dissertation, Dept. of Political Science, MIT, Cambridge, MA, Sept. 1994.

[4]*Defense Acquisition Programs: Status of Selected Programs,* General Accounting Office, GAO/NSIAD-90-159, June 1990.

[5]*Aircraft Development: Navy's Participation in the Air Force's Advanced Tactical Fighter Program,* General Accounting Office, GAO/NSIAD-90-54, March 1990.

[6]Sweetman, B., *YF-22 and YF-23 Advanced Tactical Fighters: Stealth, Speed and Agility for Air Superiority,* Motorbooks International, Osceola, WI, 1991.

[7]Abrams, R., and Miller, J., *Lockheed (General Dynamics/Boeing) F-22,* Aerofax Extra, No. 5, 1992.

[8]*Jane's All The World's Aircraft,* 1996–1997 ed., Jane's Information Group, Surrey, UK, 1996.

Chapter 8

CONCLUSION

According to the current Operational Requirements Document, the F-22 is expected to be the United States's frontline air superiority fighter for the "first quarter" of the 21st century, which is nearly 30 years from now. Its predecessor, the F-15, will likewise have filled that role for 30 years—1975 through 2004—by the time the F-22 enters service.

This is a tremendously long period of time in the context of technological advances. For example, all of the following advances have taken place over periods of 30 years or fewer:

- *Fighter aircraft, 1945–1975:* The P-51 Mustang to the F-15.

- *Fighter aircraft, 1935–1965:* The last biplane fighters to the F-4 Phantom.

- *Fighter aircraft armament, 1935–1965:* Rifle-caliber machine gun armament to radar- and infrared-guided air-to-air missiles.

- *Airborne radar, 1942–1972:* Ranges of a few miles for search only to over 100 miles with automated search/track and look-down/shoot-down capabilities.

- *Surface-to-air missiles, 1945–1975:* None operational to fully integrated air defense systems (IADS) as demonstrated in 1973 Arab–Israeli war.

- *Commercial electronics, 1967–1997:* Vacuum tubes, to junction transistors, to field effect transistors (FETs), to integrated circuits (many FETs on a single chip), with continuing improvements in processing capability (for any given size) of an order of magnitude every few years.

These dramatic developments show that 30 years is too long to be able to accurately forecast the potential advances in threat capabilities—surface-to-air missiles, air to-air missiles, manned and unmanned aircraft, and even new types of weapons. The F-22 *will* face threats that we cannot accurately foresee at this time and must therefore have a ro-

bust set of capabilities if it is to achieve its goal of providing U.S. air
dominance well into the next century.

REQUIREMENTS EVOLUTION

The specific characteristics necessary to achieve this objective have
evolved gradually and methodically since the ATF was first conceived.
Even in the 1970s, when the ATF was still thought of primarily as an
air-to-ground strike aircraft, many of the key characteristics emerged
that would prove equally applicable to the next-generation air superiority fighter:

1) Enhanced survivability—This characteristic would be needed by
   any tactical aircraft operating in the threat environment of the
   1990s and beyond.
2) Integrated avionics—During the Vietnam war, it became apparent that fighter aircraft had grown too complex for the flight crew
   to effectively manage all of the individual systems. Integration
   and automation would be needed to provide the flight crew with
   situational awareness and the ability to perform offensive and/or
   defensive actions quickly and effectively.
3) Improved reliability, maintainability, and supportability
   (RM&S)—Increasing complexity had also led to spares short-
   ages, low readiness levels, and poor mission reliability. The next-
   generation fighter would have to have a smaller logistics footprint
   and achieve higher reliability with less maintenance, without sac-
   rificing operational performance.
4) Increased range—This would provide the maximum range of basing and target options in the primary Central European scenario
   and also the ability to deploy and operate in more geographically
   dispersed regions. Although the Central European scenario was
   considered the most critical during the early stages of the ATF
   program, it was also recognized from the beginning that actual
   conflict was more likely to occur elsewhere.

The 1981 ATF Request for Information (RFI) placed equal emphasis
on air-to-air and air-to-ground. As a result of the RFI results, threat developments, and requirements dialogue within the Air Force, the ATF's
primary role was firmly defined as air superiority in 1982.

This role entails more than just air combat performance in the traditional sense, although traditional air combat performance is certainly

required. To bring its combat capabilities to the place they are needed at the time they are needed, the ATF must have longer range and greater persistence at high speeds than its predecessors. It must be highly survivable in order to control the airspace over hostile territory. The ATF must also provide its flight crew with "information dominance," achieved through a combination of its own information gathering capabilities (sensors and communications); the ability to present information to its flight crew in a logical, integrated format; and the ability to deny information to the enemy (low observables). All of these characteristics have become known as "air dominance," or the ability to dominate the airspace at any time and place of the U.S. or allied theater commander's choosing. This has been the consistent aim of the ATF since 1982, as illustrated by the following quotations:

Tactical Air Command (TAC), 1982

Air superiority is a pre-requisite to any effective air-to-ground tactical air operations.... ATF aircraft cannot avoid the conflict but must go where the targets are and beat the enemy where he is found.[1]

Commanders of the Tactical Air Forces, 1985

Our growing investment in air-to-ground capability, the need to make that investment survivable, and our important role in support of air-land battle scenarios emphasize even more the importance of providing effective air superiority at times and places of our choosing, hence ATF.[2]

Operational Requirements Document, 1996

It is imperative that the F-22 possess a first-look/first-kill advantage in all environments.... All relevant indicators used to measure the requirement for and effectiveness of tactical air power... point toward the need for an advanced air superiority aircraft. This need was originally documented in the Mission Element Need Statement (MENS) for New Fighter Aircraft, November 1981. The F-22 responds to this need by providing the capability to effectively control the air environment—enabling ground, air, and sea forces the freedom of action to conduct operations against the enemy. From the inception of the battle, F-22's primary objective will be to establish air superiority through the conduct of counter air operations in minimal time with minimal losses.[3]

The overall thrust of ATF requirements in the areas of performance, avionics, and low observables is discussed briefly next.

OVERALL PERFORMANCE

The F-22 must be able to fight and win against any adversary in any air combat regime (i.e., from beyond visual range to close-in). ATF performance objectives were initially aimed at achieving supremacy over the emerging Soviet Su-27 and MiG-29. These fighters threatened to match roughly the performance of the F-15 and F-16, and the United States has never willingly settled for parity. In the early stages of ATF development, there was also speculation that a subsequent generation of Soviet fighters, perhaps with supercruise capability and/or low observables, would follow the Su-27 and MiG-29. These particular threats never materialized. However, the Su-27 and MiG-29 have been extensively developed and upgraded. Su-27 variants have been equipped with canards and thrust vectoring for improved maneuverability. Both fighters continue to be upgraded with advanced weapons and targeting systems and are available to international customers. New Western European fighters, with performance superior in some respects to the F-16 and/or F-15 along with advanced weapons, will also be entering the export market. It is possible that other aircraft, not yet in development, will be entering service while the F-22 is still operational.

During the course of the ATF/F-22 program to date, its requirements have been continually refined, with affordability in mind, as better information on achievable ATF performance, and threat capabilities, have become available. Various requirements with high cost or risk, and minimal value, have been relaxed as much as possible while still maintaining a margin of superiority over any existing or projected threat. In summary, the F-22's performance requirements represent a finely balanced set of characteristics intended to provide the greatest possible assurance of U.S. air dominance for the first quarter of the 21st century.

AVIONICS

The F-22's avionics requirements have been carefully developed to achieve a high degree of synergy with the air vehicle performance and signature characteristics. This entails a high degree of integration to provide the pilot with useful information rather than overwhelming quantities of data. Another important feature is the ability to operate in both active and passive modes with strong attention to controlling electronic emission (EMCOM). This ability, combined with the F-22's speed and stealth, will allow the F-22 pilot to actively gather information and then act quickly and decisively while denying information to the enemy.

Signature reduction was first considered for the ATF during the air-to-surface period of the 1970s, as part of an overall emphasis on survivability. However, the ATF program proceeded independently of "black world" low-observable (LO) programs approximately through the completion of the 1981–1982 RFI effort. The effectiveness of LO—how much signature reduction could be achieved, at how much penalty to aerodynamic performance, and what the operational benefits would be—were therefore relatively unknown quantities in the ATF program up to that time.

For example, a White Paper on the ATF prepared by USAF/RDQ in 1982 included a requirement for a frontal radar cross section (RCS) of 0.1 m^2. In reviewing the White Paper, TAC concurred with the level but suggested that the requirement be extended to cover the beam and aft quadrants.[4] Several aircraft that had flown by that time, including the Have Blue and Tacit Blue demonstrators and the F-117A strike aircraft, had proven that much lower signatures were achievable. The resulting impact on air warfare would be revolutionary.

In late 1982 or early 1983, the ATF program was directed to place increased emphasis on low observables to insure that various low observable technologies were properly assessed in an integrated manner. The concept development RFP was modified (after release) to require the contractors to present LO trade study data. ASD/XRJ, the Directorate of Low Observables that managed the various LO programs, became involved with ATF, and security arrangements were set up with each ATF contractor to handle the exchange of information relating to LO.

During the course of the concept exploration phase, signature goals were developed for the ATF based on the contractor studies and input from signature experts in XRJ. The signature levels were generally what would be considered "very low observable" (VLO), except in the aft sector because of the perceived performance penalties of LO engine nozzle treatments. These goals were contained in the Dem/Val RFP.

However, after RFP release, the program office again came under pressure (from OSD) to strengthen the emphasis on LO, particularly in the aft sector. There was some feeling that the technology was mature enough to achieve low aft aspect signatures and still allow for high aerodynamic and propulsive performance, but that it would not be accomplished unless the RFP asked for it. Another change was to require full-scale RCS pole model testing during Dem/Val. The demonstrations proved that the desired signature levels could be achieved, and so the initial Dem/Val LO requirements have been maintained without major change into the EMD phase.

The F-22 is reportedly showing every indication that it will meet its signature requirements. There was a problem discovered in late 1993, relating primarily to the signature contributions of various details. Signature performance was recovered through a year-long effort involving redesign of the control surface and nozzle seals, reduction in the number of apexes on actuated doors and access panels, a large reduction in the number of drain holes on the aircraft, and other similar detail changes.

EXTERNAL PRESSURES

The ATF program often came under pressure to give greater attention to various considerations, in particular to one or more of the following: 1) RM&S, 2) affordability, 3) alternative solutions, and 4) technical risk/technology availability (particularly avionics). This was usually a result of some outside individual or organization simply being unaware of how much attention the ATF program was already devoting to the subject in question.

RELIABILITY, MAINTAINABILITY, AND SUPPORTABILITY

Improved RM&S has consistently been a high priority in the ATF program. The statement of need that was validated in November 1984, preparatory to Milestone I, contained quantitative RM&S requirements, which represented, in general, a factor-of-two improvement relative to F-15 RM&S levels. These requirements have undergone only minor adjustments through the present time.

Examples of emphasis on RM&S in the ATF program include the following:

1) RM&S-related evaluation criteria were included in the analysis of the contractor concept development studies.[5]
2) Affordability and RM&S were ranked coequal with system performance, as evaluation factors for the Dem/Val RFP.
3) The Dem/Val phase included many supportability demonstrations (e.g., 400 individual demonstrations by the YF-22 team).
4) Many supporting technology programs were oriented toward improved RM&S (for example, Ultra Reliable Radar).
5) Improved RM&S is one of the key benefits of the overall integrated, modular avionics architecture.
6) The Joint Advanced Fighter Engine (JAFE) engine program was

aimed at designing improved RM&S into the engines from the start, by reducing the parts count, eliminating maintenance nuisances such as safety wire, reducing special-use tools, using common fasteners, improving durability, improving the diagnostics, etc.

Two key benefits of improved RM&S are increased readiness and reduced life cycle cost (LCC). The relative importance of these two factors has shifted since the ATF program was initiated. At the beginning, the emphasis on RM&S was driven by need for high readiness and sortie generation rates in a high-intensity conflict, and in the face of enemy attacks on friendly bases (including, potentially, maintenance facilities). Problems with low readiness levels in advanced systems such as the F-14 and F-15 were high-visibility issues during that period. At present it is considered unlikely that the United States will be involved in anything as desperate as a NATO–Warsaw Pact confrontation in Europe. However, the need for high readiness with minimum logistic support is still a high priority, as we strive to control LCCs while meeting worldwide military commitments with downsized forces.

AFFORDABILITY

The ATF program approach to affordability, from the beginning, has been an overall systems engineering process to arrive at the most cost effective weapon system. Attention has focused both on initial acquisition cost and on LCC. Aircraft weight has been important because it is one of the most reliable indicators of probable cost early in the design process.

The conceptual designs submitted in response to the 1981 RFI (as well as in-house ASD analyses) indicated a likely takeoff gross weight of approximately 55,000 lb for an advanced air superiority fighter with supersonic cruise and maneuvering capability.[6] This estimate remained essentially unchanged during the funded concept development investigation of 1983–1984 with several contractor design concepts weighing as much as 60,000 lb. Cost estimates at that time indicated a likely unit cost between $40 and $45 million (FY85 dollars) for an aircraft in the 53,000–55,000 lb region.[7]

An important goal, going into Dem/Val, was to identify the correct combination of technologies, design characteristics, and requirements tradeoffs to make the system as affordable as possible while maintaining its essential mission capabilities. Whereas low cost and low weight are both desirable, it was recognized that setting hard limits prematurely could be counterproductive. It was noted, for example, that when too much emphasis is placed on initial acquisition cost early in a program, one of the first things to be compromised is RM&S.[5,8]

However, at a meeting of the Air Force Systems Acquisition Review Council (AFSARC) in November 1984, just prior to the planned Milestone I review, a $40 million cost goal was imposed, along with a 50,000-lb limit on takeoff gross weight (TOGW). Concerns over affordability delayed the actual Milestone I review by nearly a year. At a subsequent AFSARC review in August 1985, the cost goal was further reduced to $35 million (FY85 dollars, assuming 750 aircraft produced at a rate of 72 per year), while the 50,000-lb weight limit was maintained.

There was a strong effort early in Dem/Val to make the necessary requirements trades to achieve the weight and cost goals. Aerodynamic performance requirements were "scrubbed" thoroughly to arrive at a balanced set of parameters that would ensure superiority over any existing or projected threat while eliminating those requirements that were excessively difficult to meet but of only marginal value. Top speed requirements were adjusted to avoid the cost and risk associated with variable-geometry inlets. Avionics requirements were looked at critically, and some requirements that had been included in the original 1984 Statement of Operational Need, and carried forward into the operational requirements document (first released in 1987), were eliminated in 1988–1989. The requirement for an infrared search and track (IRST) system was deleted. The required radar field-of-regard was reduced to a value that could be achieved without side-looking radar arrays. (However, growth provision in both of these areas has been maintained.) The requirement for thrust reversing was deleted. All of these trades reduced cost, weight, and technical risk. At the same time, the strong RM&S requirements were kept essentially unchanged.

However, by 1988 or 1989, it was finally accepted that the essential mission capabilities of the ATF could not be achieved within the somewhat arbitrary cost and weight caps that had been established before many of the necessary design, technology, and requirements trades were accomplished. Sherman Mullin, general manager of the winning ATF team during the Dem/Val phase, recalled:

> By the time we completed the formal System Design Review in November 1989 it was clear that the convergence of ATF performance, weight, and cost was being achieved and that the specific system engineering approach we had implemented was really working. We never did find the magical 50,000 lb, $35 million airplane, but we found the "right" answer.[9]

In fact, Lockheed's winning ATF design reportedly had a higher gross weight than Northrop's, due in part to the use of thrust vectoring on the Lockheed aircraft as well as having four tail surfaces instead of the two used in Northrop's design.[10] However, by the time the EMD source selection was made, the designs were sufficiently mature that weight did not have to be relied on as the best indicator of probable cost.

There is not a formal specification for takeoff gross weight of the F-22 in EMD. A specification for empty weight was established at the beginning of EMD, but this has never been in danger. The specified empty weight is sufficiently conservative that, if the F-22 experiences weight growth, it will fail to meet its maneuverability requirements before it will exceed the specified empty weight.[11] Furthermore, weight is not listed as a requirement in the Operational Requirements Document, it is not defined as a critical system characteristic, and it is not a key parameter in the acquisition program baseline.[12] This arrangement correctly places emphasis on the F-22's performance rather than on weight, which, while important, is just one of several factors that directly affect performance.

The contractor team has kept a high level of emphasis on weight control. Lockheed Program Manager Mickey Blackwell stated, "We've got to declare a crisis to avoid a crisis," implying that a strong effort to control weight was not indicative of a problem, but of the solution.[10] A small amount of additional weight growth occurred during early EMD. Since then, weight estimates have been relatively stable.

As already noted, the unit flyaway cost goal going into Dem/Val was $35 million in fiscal year 1985 dollars, which corresponds to $41.2 million in fiscal year 1990 dollars. Following the Major Aircraft Review (MAR) of 1990, the planned production rate was reduced from 72 to 48 per year, and the production start was slipped four years to 1996. As a result of these changes, together with more detailed estimates based on greater definition of the designs and requirements, the projected average unit flyaway cost increased to $51.2 million fiscal year 1990 dollars. At Milestone II, the Defense Acquisition Board reduced the production quantity to 648 (from 750). This resulted in a further increase of the average unit flyaway cost, to $56.9 million fiscal year 1990 dollars. Thus the total increase during Dem/Val, relative to the initial goal, was $15.7 million fiscal year 1990 dollars. Approximately $5.2 million of the increase is attributable to technical reasons, whereas the remainder is due to schedule slips and reductions in the production run. A further increase to $61.2 million in fiscal year 1990 dollars occurred following the

Bottom-Up Review in 1993 when the production run was further reduced to 442.* The effects of the slips on unit cost, when expressed in then-year dollars, are even greater because the effects of inflation must then be included.[13]

The General Accounting Office reported in 1991 that over 80% of ATF program cost increases to date were the result of schedule slips, decreases in production rate, and the effects of inflation.[14] Cost increases since then have been similarly driven by externally imposed quantity reductions, schedule slips, and inflation.[13]

The history of ATF unit cost estimates, from Dem/Val start until immediately after the Bottom-Up Review, is shown in Table 10. In addition to unit flyaway cost (UFC), the table shows unit procurement cost (UPC) and program acquisition unit cost (PAUC). UFC is defined as the cost of recurring hardware, software, and management (i.e., the total cost associated with producing and delivering each ATF), plus nonrecurring start-up and allowance for changes required to procure the aircraft. UPC is defined as the total procurement funding divided by the total buy. UPC therefore includes the UFC, plus technical data, publications, support and user training equipment, factory training, and initial spares. PAUC is defined as the total program cost divided by the total production buy, including RDT&E and military construction (MILCON) costs as well as procurement costs. The far right column in the table indicates the start and end dates of the procurement schedule.[13]

It must be remembered that, "while the $35 million target was well known in the public arena, it was never a program baseline. This target was created to balance performance with affordability in contractor design proposals."[13] The ATF program has placed strong emphasis on affordability, from its inception. However, it has not been a conception of affordability that is based simplistically on arbitrary flyaway cost and weight limits. Rather, the program has aimed to arrive at the most *cost-effective* solution for achieving U.S. air dominance for the first quarter of the 21st century. All discussions of affordability must be based on an F-22 that is the most capable air superiority fighter in the world during that period. If it is only second-best, then it is not affordable no matter how little it may cost.

---

* Planned F-22 production quantity was again reduced to 338 aircraft following the 1996 Quadrennial Defense Review.

Table 10   History of projected Advanced Tactical Fighter unit costs

Event	Date	UFC, $M			UPC, $M			PAUC, $M			Total acft.	Peak rate	Proc. sched.
		FY90$	FY96$	TY$	FY90$	FY96$	TY$	FY90$	FY96$	TY$			
D/V Start	Oct. 86	41.2	48.1	51.6	54.0	63.0	67.5	69.1	80.6	85.7	750	72	92-05
MAR	Aug. 90	51.2	59.8	81.2	65.8	76.8	104.3	85.5	99.8	136.0	750	48	96-14
MSII DAB	July 91	56.9	66.4	103.2	67.1	78.3	122.4	92.7	108.2	152.5	648	48	96-12
BUR	Sept. 95	61.2	71.4	101.2	73.7	86.0	121.7	112.3	131.1	166.5	442	48	97-14

F-15 and F-16 derivatives have been proposed repeatedly as cheaper alternatives to the ATF. However, the ATF offers three unique capabilities that are essential to the effective performance of its mission. These are high performance integrated avionics, supersonic cruise, and low observables. Avionics with capabilities comparable to those on the ATF could conceivably be retrofit on an existing aircraft design; however, they would be just as expensive on any other aircraft as they will be on the ATF. Some measure of supercruise performance might also be achieved by putting an ATF engine into a modified F-15 or F-16. Avionics are expected to represent approximately 40% of the unit fly-away cost of the ATF, while its two engines account for a further 20%. Therefore, with ATF avionics and engines, the "cheaper" derivative aircraft is now carrying about 60% of the cost of an ATF (or 50%, if it only has one engine), *not* counting the cost of its own airframe. And it does not yet have the last critical characteristic: low observables.

Sometimes claims are made that low observables can also be "added" to an existing fighter design. For example, the following statements were made shortly after the 1990 Major Aircraft Review:

> One senior Pentagon official has noted that the Falcon 21 could provide real competition to the ATF, since the ATF stores all its weapons internally, limiting its payload. The semi-submerged [i.e., conformal] weapons carriage configuration of the Falcon 21 would give it more flexibility in weapons carriage, while still providing a degree of low observability...."[15]

Such claims ignore some basic physics. A "degree of low observability" is of little value. According to Sherman Mullin, who served as Lockheed's F-117A program manager prior to his involvement in the ATF program, radar cross section

> ...is the only term in the [radar detection range] equation under the control of the aircraft or missile designer. All other parameters are controlled by the radar systems designer and the laws of physics. This is an important and enduring fact of life, not well understood in certain quarters...where stunningly ignorant comments about stealth are frequently heard.... Maxwell's equations are not a passing fancy.[16]

The central problem is that detection range does not vary linearly with radar cross section (RCS), but rather with the fourth root of RCS. This is a very weak dependence, requiring very large changes in RCS to produce moderate reductions in detection range. For example, reducing the RCS by a factor of 10 produces only a 44% reduction in the range at which an aircraft can be detected by a given radar. Thus, if one aircraft can be detected at 100 miles, another aircraft with 1/10 the RCS can be detected at 56 miles, which still results in considerable exposure to enemy air defenses. To reduce the detection range of the same hypothetical radar system to 10 miles, the RCS of the aircraft would have to be reduced by a factor of 10,000.

Such extremely low values of RCS can in general only be achieved by designing a system for low RCS from the ground up. A conventional aircraft simply has too many features that reflect radar waves, including not only its surfaces, edges, and corners, but also the sensor apertures, air data probes, ram air cooling inlets, communication and ECM antennas, etc., many of which items *individually* exceed the *total* RCS that must be achieved to be effectively "stealthy." Thus, while appealing at first glance from a financial point of view, upgraded versions of existing aircraft would never be able to match the effectiveness of the F-22 as long as low observables remain an important consideration; and so far, low observables have proven to be the most effective—or even the only effective—way to survivably penetrate heavily defended hostile airspace. Such derivative aircraft typically provide one-third of the capability at two-thirds of the cost of the ATF. Nevertheless, pressure to consider cheaper alternatives to the ATF has continued and will probably continue until the last F-22 comes off the production line.

TECHNICAL RISK

The ATF/F-22 program has often been criticized for carrying excessive technical risks, by individuals or organizations who do not realize or appreciate the extent of effort that has been devoted to risk reduction and to making sure that the necessary technologies are mature and available. The Milestone Zero decision in November 1981 specifically endorsed a strategy of "focused technology thrusts" rather than early prototyping of specific aircraft designs. Even the program name during the conceptual phase was Advanced Tactical Fighter Technologies. The program management directive (PMD) that implemented the Milestone Zero decision stated the following:

> The scope of the program is changed to focus on technologies and concept development of a new generation of tactical fighters to be produced in the 1990s and to add development of an advanced technology fighter engine. The program title is changed to Advanced Tactical Fighter (ATF) Technologies with two projects entitled Advanced Tactical Fighter and Joint Fighter Engine.

...This program will be structured to develop, design, and support options to meet both air-to-air and air-to-surface manned fighter mission needs and to insure that the technical base to support full scale development, production, and combat support of these design options is available. [The Advanced Tactical Fighter project] will develop aircraft concepts, designs and support alternatives and will validate technical approaches for the primary mission of air superiority with a secondary air-to-surface role. [The Joint Fighter Engine project] will develop the critical advanced propulsion system technologies and support concepts and will validate them in a baseline technology demonstration fighter engine. The ATF program will also provide timing and capability focus for technology base development activity in avionics, weapons and weapons integration.[17]

The strong emphasis on supporting technologies was continued in the Dem/Val phase. Although ATF prototypes were incorporated into the Dem/Val phase in 1986 (largely by outside direction), the focus remained on risk reduction, and rather than using the prototypes as a competitive evaluation tool, each contractor team was encouraged to use the prototype flight-testing in whatever ways would be of greatest benefit to their overall risk reduction efforts.

The success of all four Dem/Val prototype airframe/engine combinations in achieving unprecedented performance characteristics is just one example of how effectively technical risk was managed in the ATF program. The supporting technologies received just as much attention (and funding) as the airframes, in fact, considerably more funding during the concept development phase. This is particularly true of avionics, which was identified by all of the bidders for the Dem/Val phase as the single highest technical risk area. The relative funding devoted to airframes, avionics, and propulsion during the first two phases of the ATF program is illustrated in Fig. 94. In both phases the airframe received the least funding of the three categories.*

It can be seen that the risks inherent in achieving the ATF's unique capabilities were appreciated from the beginning, and a comprehensive effort was made to manage those risks. The key concept is manage, rather than avoid, because simply avoiding risks does not lead to any

* Note that this only includes the specific efforts noted (e.g., INEWS and ICNIA), which were directly related to the ATF program and for which cost figures were readily available. Extensive additional funding was devoted to technology base efforts in all three areas, with the greatest funding probably being devoted to the avionics technology base. For example, it is known that Galium Arsenide (GaAs) technology, which was just one of several technology areas contributing to the ATF's advanced radar capabilities, received $78 million in Air Force funding between 1970 and 1988.

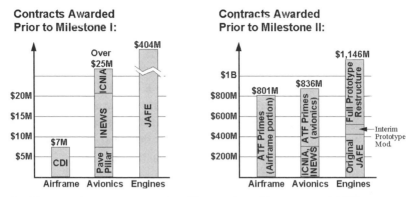

Fig. 94   Relative emphasis during CDI and Dem/Val phases of ATF program.

advancement in capabilities. This point is well stated in the following remarks by Sherman Mullin, general manager of the winning ATF team during Dem/Val:

> Aeronautical engineering was not meant to be a profession for those engineers obsessed with very low risk.... Now we have risk averse types of engineers in abundance.... In the case of the F-22 we knowingly took on an enormous system integration challenge to combine supercruise, low observability, integrated avionics and classic fighter performance in a new airplane.... There is no such thing as a low risk future program in advanced aeronautics. Yet this oxymoronic phrase is heard with increasing frequency. We must continue [to accept technical challenges and a certain amount of risk], or we will slowly whither away.[9]

In addition to managing its own technical risk, the ATF/F-22 program paved the way for many other programs. For example, it pioneered the common modular, integrated avionics approach being implemented in Joint Strike Fighter and the upgrades to other platforms, and in many ways it preceded many commercial avionics developments, both in architecture and functionality.

FINAL OBSERVATIONS

The F-22 is the product of a broad-based, comprehensive effort to develop a weapon system that will provide air superiority against any adversary at the time and place of the U.S. or allied/coalition theater commander's choosing. Over a period of 15 years, technologies have been matured, demonstrated, and transitioned; airframe and engine designs have been developed and demonstrated; and requirements have been carefully refined to provide this capability as cost effectively as possible.

The world has changed since these efforts were initiated. However, with major weapon system development cycles routinely exceeding 15 or sometimes 20 years, we must accept the fact that the world often will change significantly during the time that a major weapon system is developed. The weapon systems under development at any given time therefore will seldom, if ever, be exactly tailored to the latest projections of what our forces will need. If we insisted on always having such a match, we would never develop anything. We would be constantly changing the requirements, or canceling programs altogether and starting over.

We must give the warfighters the best systems possible, given the knowledge that is available at the time critical decisions are made in the development of those systems. Then, *to some extent,* we must depend on the warfighters to make the best use of those weapon systems in situations that will be different from those envisioned when the systems were under development.

The ATF has been designed to perform a mission that transcends any specific geopolitical situation. Regardless of the location or of the identity of the adversary, air superiority is crucial to accomplishing military objectives rapidly and with minimal casualties. The unique capabilities of the F-22—its ability to cover large distances at supersonic speed and with minimal probability of being detected; avionics that will give the pilot "information dominance," and the ability to defeat any threat when the pilot chooses to engage—are all well suited to meeting the diverse military commitments that the United States will face over the next 30 years.

REFERENCES

[1]Nordmeyer, Capt. USAF, TAC/DRFG, to TAC/CC, /CV, and /CS, Staff Summary Sheet: *Advanced Tactical Fighters (ATF),* May 17, 1982.

[2]Bazley, Gen. USAF, Donnelly, Gen., USAF, and Kelley, Lt. Gen. USAF, Memorandum to Gen. Gabriel, May 14, 1985. SECRET. (Unclassified information only used from this source.)

[3]Operational Requirements Document, *F-22 Advanced Tactical Fighter.* SECRET. (Unclassified information only used from this source.)

[4]*White Paper on Advanced Fighters,* USAF/RDQT, with TAC/DRFG Comments, early 1982 (originally SECRET; declassified Jan. 10, 1997).

[5]Ferguson, P. C., *Oral History Interview: Brigadier General Claude M. Bolton, Jr.,* Office of History, Aeronautical Systems Center, Wright–Patterson AFB, OH, Aug. 13, 1996.

[6]Ferguson, P. C., *From Advanced Tactical Fighter (ATF) to F-22, Part I—To Milestone 0 and Beyond: 1970–1982,* Office of History, Aeronautical Systems Center, Wright–Patterson AFB, OH, May 1996.

[7]Various Weekly Activity Reports (WARs), from ATF System Program Office to the Commander, Aeronautical Systems Div., Nov. 23, 1983, Nov. 13–21, 1984, Dec. 6–12, 1984, Dec. 26, 1984–Jan. 3, 1985, May 29–June 5, 1985, and June 13–19, 1985.

[8]Wade, Assistant Secretary of Defense for Acquisition and Logistics (SAF/AL), Comments at ATF R&M Briefing Nov. 4, 1985, quoted in WAR, Oct.–Nov. 6, 1985.

[9]Mullin, S. N., "The Evolution of the F-22 Advanced Tactical Fighter," Wright Brothers Lecture, AIAA Paper 92-4188, Washington, DC, Aug. 24, 1992.

[10]Twigg, J. L., *To Fly and Fight: Norms, Institutions, and Fighter Aircraft Procurement in the United States, Russia, and Japan,* Ph.D. Dissertation, Dept. of Political Science, MIT, Cambridge, MA, Sept. 1994.

[11]Graves, J. T., Deputy Director of F-22 SPO, "Acquisition Executive Monthly Report (AEMR) for the F-22 (ATF) Program," Sept. 30, 1991.

[12]Snodgrass, M. A., Maj. USAF, *Information Paper: F-22 Weight Reduction Efforts,* SAF/AQPF, Oct. 22, 1993.

[13]Walley, R. A., Maj. USAF, *Information Paper: History of F-22 Unit Cost Estimates,* Office of the Assistant Secretary of the Air Force (Acquisition), Fighter Div. (SAF/AQPF), March 21, 1996.

[14]*Reasons for Recent Cost Growth in the Advanced Tactical Fighter Program,* General Accounting Office, GAO/NSIAD-91-138, February 1991.

[15]Brown, D. A., "General Dynamics Evaluates Concepts for F-16 Successor," *Aviation Week and Space Technology,* June 11, 1990, pp. 21–22.

[16]Mullin, S. N., Private communication with David Aronstein, Aug. 17, 1996.

[17]Program Management Directive No. R-Q 7036(5)/63230F, *Advanced Tactical Fighter (ATF) Technologies,* USAF/RDQT, Aug. 24, 1982.

Appendix

Have Blue Technology Demonstrator and Aircraft Signature Reduction

Introduction

By the early 1970s, in the wake of the Southeast Asia War, it had become obvious that offensive air operations were being severely jeopardized by radar-controlled air defense weapons. The reality was that modern combat aircraft had essentially the same basic susceptibility to being detected, tracked, and engaged by radar as did their World War II ancestors; however, the radar-directed threats had become vastly more lethal and were reaching the stage of being potentially devastating. This reality was reinforced by the dramatic impact of recently introduced Soviet SAMs on the combat attrition of U.S.–supplied aircraft during the 1973 Arab–Israeli War. Self-protection jamming systems, added to tactical aircraft during the Vietnam War, often could not adequately deal with such threats. It became clear that major reductions in aircraft radar signature had to be made if the manned strike aircraft was to remain a viable weapon system. This was the genesis of the Have Blue program. Have Blue was created to demonstrate and validate the technical feasibility of employing high-performance aircraft having very low radar and infrared signatures to neutralize enemy threat radar systems; its success led directly to the F-117A.

Have Blue Program

In late 1974, the Advanced Research Projects Agency (ARPA) initiated an effort to develop an aircraft with very low radar and infrared observability. Five companies participated in the initial studies, with Lockheed entering the effort later. By August 1975, the field had been narrowed to McDonnell Douglas, Northrop, and Lockheed. Proposals were sought from these companies for the design and development of a low-observable testbed aircraft, the XST (Experimental Survivable Testbed).

On November 1, 1975, Lockheed and Northrop were selected for Phase I. This phase consisted of preliminary design and analysis; con-

259

struction and radar signature testing of one-third scale and full-scale models; low-speed, high-speed, and propulsion (inlet) wind-tunnel tests; and initial flight control simulations. Lockheed was selected as the winner of Phase I and given a contract to proceed with Phase II in April 1976. This would consist of the design, construction, and flight-testing of two demonstrator aircraft. First flight was scheduled for December 1977. Increased security was applied, management passed to the Air Force, and the name Have Blue adopted. The initial contract covered a 10-month flight-test program, which occurred from December 1977 through September 1978. Flight testing after October 1, 1978, was described as "follow-on testing" and was funded under a new contract.[1]

The Have Blue aircraft were not intended to be prototypes but rather were advanced concept technology demonstrators with three specific objectives:

1) Validate the achievement of low observability including very low radar cross section (RCS) and infrared signature in a real aircraft.
2) Demonstrate that such an aircraft could have acceptable handling qualities and performance.
3) Demonstrate capabilities of modeling and test techniques to accurately predict low-observable characteristics of actual aircraft in flight.

Two Have Blue aircraft were built, of minimum size and cost. They featured a unique "faceted" low-observable design concept that had been developed (and was later patented) by Lockheed. This design approach was consistent with the analytical techniques for predicting radar signatures that were available at the time of their design. These techniques could not adequately handle curved surfaces and other complex shapes.

Lockheed's formerly classified patent application[2] titled "Vehicle," filed with the commissioner of U.S. Patents and Trademark Office on February 11, 1979, provides fascinating insights into the faceted aircraft design concept. In the application Lockheed noted that radar detectability is an inherent characteristic of a vehicle's shape and that previous attempts to reduce radar signature by addition of anti-reflective coatings had not been successful in significantly reducing detectability. The invention was described in the patent application as

...a vehicle whose external surfaces are configured to make such vehicles *substantially invisible** to radar by reducing the signal received below receiver sensitivity levels and/or clutter.... It is a further object of the present invention to provide a vehicle having a substantial absence of curved surfaces....the desired stealth capability (i.e., low radar cross section) is imparted to the vehicle ...through the use of a basic polyhedron shape, the respective surfaces of the vehicle being planar facets.... These facets are arranged so as to present the illuminating source with high angles of incidence, thus causing the primary reflected power to be in a direction of forward scatter, i.e., away from the source.... Facets and edges are also sometimes constructed partially or totally from, or are treated with, antireflective materials and surface current density control materials.... The flat, facet surfaces concentrate scattered energy primarily into a forward scatter direction, minimizing side lobe direction magnitudes.... Thus, the tracking radar receives either small undetectable signals or only intermittent signals which interrupt continuous location and tracking ability.... The desired characteristics may be provided while also maintaining reasonable and adequate aerodynamic efficiency *in the case of an airborne vehicle* (Ref. 2).[†]

The illustrations in Figs. A1 and A2 show a Have Blue–type aircraft configuration. They were included in the original Lockheed patent application. Noteworthy are the highly swept wings (72.5 deg), the complete faceting of all airframe surfaces, the gridded engine inlets, and the unique nozzle design. The latter features were to prevent radar energy from reaching the engine front and rear faces. The numbers on the drawings were referenced in the text of the application and were used to describe specific aircraft design features.

Design operational load factors for the Have Blue aircraft were only $+3\,g$ and $-1\,g$ with an airspeed envelope that ranged from about 150 to 500 kn; however, although modest, these parameters were consistent with their role. Gross weight was 12,000 lb and wingspan was only 22 ft, 6 in., much smaller than any potential operational stealth aircraft. The Have Blue aircraft have often been referred to as subscale prototypes for the F-117A, either 60% (by linear measure) or 25% (by

---

* Emphasis added. Development of a practical combat aircraft having *complete* invisibility to all threat radars at all possible operating frequencies is probably not feasible even with aggressive application of signature reduction techniques.

† Emphasis added. The faceted low-observable concept was also used on another (nonaircraft) Lockheed vehicle design, the U.S. Navy experimental stealth surface ship known as Sea Shadow, which was designed and built by the Skunk Works during the same period that the F-117A was in production.

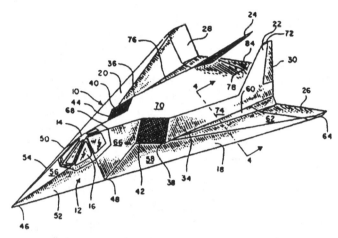

Fig. A1 Upper front quarter view of the Lockheed faceted aircraft design concept.[2] The vertical fins were swept back 35 deg and canted inward 30 deg. The Have Blue aircraft used all moving rudders, as shown in Fig. A3, rather than the fin/rudder arrangement seen here.

weight), but this is not really the case as the size, payload, and configuration of an operational stealth aircraft had not yet been decided at the time the Have Blue aircraft were built. It would be more accurate to say that the F-117A was adapted from the Have Blue design concept.[1] The Lockheed Have Blue and the F-117A aircraft are shown to the same scale in Fig. A3.

Construction of Have Blue number 001 began in July 1976. Only the second aircraft had full signature reduction treatments. The first Have Blue would be used for handling qualities and flight performance evaluation only; the second would be used to verify the predicted low-

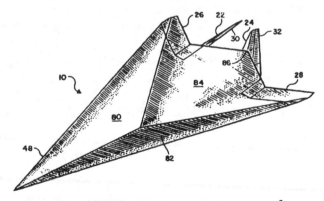

Fig. A2 Lower rear view of Lockheed's faceted design concept.[2] Note how the exhaust nozzle is hidden from view by the extended lower shelf, which had a scarf angle of 54 deg.

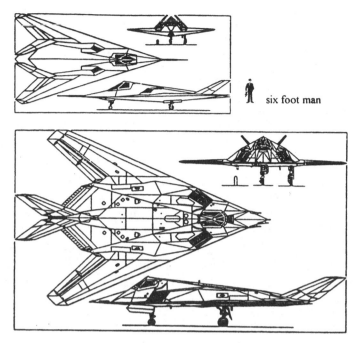

six foot man

Fig. A3   A same-scale comparison of the Lockheed Have Blue and F-117A aircraft.[1]

observables performance. In keeping with their respective purposes, Have Blue 001 was equipped with a large test instrumentation nose-boom; aircraft 002 featured a low-observable air data system that completely dispensed with external pitot tubes.

The Have Blue demonstrators were statically unstable about all axes. This was not an inevitable consequence of their faceted shape, but rather a measure to give the designers the freedom to reduce the size of the tail surfaces and/or to locate the center of gravity farther aft than would be otherwise possible, thereby reducing trim drag and increasing the maximum usable lift. "Artificial" stability was provided by a quadraplex fly-by-wire control system modified from the F-16. At the time of the Have Blue program, the F-16 was not yet operational although the control system had flown on the YF-16 Lightweight Fighter demonstrator in early 1974.

The Have Blue aircraft had a movable two-position "platypus" control surface, which was nested during high-speed flight and deflected during takeoff, landing, or flight at high angles of attack (above 12 deg). This complemented the elevons and provided improved pitch control. No mission equipment was included, no flaps or speedbrakes, and no air-to-air refueling capability. Maximum use was made of off-the-shelf components. These included the landing gear from the Northrop F-5, a modified General Dynamics F-16 flight control system as already noted, and proven nonafterburning General Electric J85-GE-4A engines from the U.S. Navy's North American Rockwell T-2C advanced trainer.

Flight-testing was performed under very tight security. Have Blue 001 made its first flight on December 1, 1977. In addition to an extremely high takeoff and landing speed, the Have Blue aircraft possessed some unusual flying qualities. Because of the extremely high sweep and small span of the wings, climb was sluggish and high sink rates could develop during landing approaches. The maximum lift-to-drag ratio (L/D_{max}) was approximately 7.5; as can be seen in Fig. A4, this was lower than that of any operational jet fighter, including Lockheed's Mach Two F-104.

From December 1977 through early 1978, Have Blue 001 flew 36 test sorties, all of them for functional checkout, flying qualities, and performance evaluation. The aircraft was lost on May 4, 1978. Airspeed had dropped too low during a landing approach. The platypus operated au-

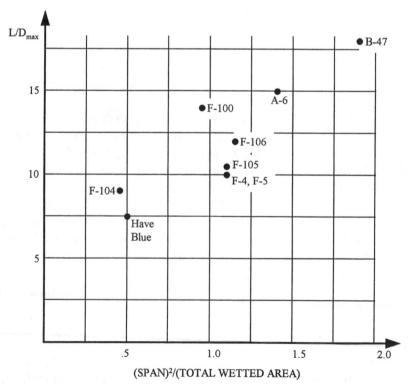

Fig. A4   Have Blue L/D_{max} compared to other military aircraft.[1]

tomatically as designed, but "pilot–platypus interaction" (pilot-induced oscillation) resulted in hard ground contact that damaged the right main landing gear. The landing attempt was aborted, and the pilot retracted the gear for a go-around. The gear would not fully extend for a subsequent landing attempt; the pilot ejected and was seriously injured.

The second Have Blue (002) first flew in mid-1978 and completed 52 sorties. Ten of these were for functional checkout and performance evaluation, and the remaining 42 were for low-observables testing. During these flights, very low radar signatures were repeatedly demonstrated and radar-directed threat systems were decisively neutralized. The second Have Blue aircraft was lost on July 11, 1979. An engine exhaust duct weld failed, allowing exhaust gases to leak into the engine compartment. This resulted in failure of one of the two hydraulic systems. The pilot lowered the platypus for return to base, but the second hydraulic system failed in a similar manner to the first one. This resulted in complete loss of aircraft control. The pilot ejected safely. When aircraft 002 was lost, the Have Blue program was within two or three sorties of planned completion.

The following sections contain a brief discussion of basic radar and infrared fundamentals as they relate to low-observable aircraft design. This is necessary to appreciate the technical achievement of the Have Blue program in vastly reducing aircraft signature.

(Very) Brief Radar Overview

An omnidirectional radio detection system was successfully demonstrated by Christian Hülsmeyer in Germany in 1903 (the same year as the Wright brothers' first powered controlled airplane flight). By the mid-1930s, radio technology had progressed to the point where experimental radars using focused beams and pulsed transmissions were able to determine both the range and direction of distant moving objects including ships and aircraft. (The acronym "radar" was developed during World War II; it stands for *radio detection and ranging*.) Radar surveillance and fire control systems were in operational service in Germany, Great Britain, and the United States by the start of World War II. They immediately had a major impact on military operations on land, at sea, and in the air and were rapidly introduced by all of the major warring nations. (A U.S.–developed SCR-270 surveillance radar tracked the Japanese naval carrier-based strike force en-route to attack the U.S. Pacific Fleet at Pearl Harbor and other Hawaiian bases on December 7, 1941.)

Radars are an integral component of all modern integrated air defense systems. They perform the vital functions of surveillance, acquisition, tracking, fire control, and missile guidance. Radars detect and track targets by using either a movable or an electronically scanned antenna to transmit and then receive reflected electromagnetic energy. Radars come in a variety of forms including pulsed, pulsed Doppler, and continuous wave. The pulsed radar determines range by pulse delay techniques (e.g., measuring the time intervals between transmission and receipt of each pulse). Pulsed Doppler radars measure range by either pulse delay or frequency modulation techniques. The continuous wave radar measures velocity and closing rates by sensing Doppler frequencies; it can measure range by modulating transmitter frequency. As with all systems, specific mission requirements determine the technical parameters (such as operating frequency) that in turn become the basis for each radar's detail design.

OPERATING FREQUENCY

A radar's operating frequency f and its wavelength λ are inversely proportional to one another and are related by the simple equation shown here:

$$f = c/\lambda \qquad \qquad (A1)$$

where λ is normally expressed in meters and c is the speed of light (approximately 3×10^8 m/s). This is the speed of propagation in free space of all forms of electromagnetic energy, including radar waves.

Selection of a radar's operating frequency depends on its intended role. Lower frequency (longer wavelength) radar emissions are less attenuated by the atmosphere and adverse weather; these radars thus provide much longer detection ranges. Therefore, surveillance radars (such as the Soviet-developed Spoon Rest and Tall King) generally operate at lower frequencies (in the range of 0.1 to 0.2 GHz). Their detection ranges against conventional aircraft are only limited by line of sight (curvature of the Earth or terrain masking). These ranges are on the order of over 250 n miles for conventional aircraft flying at typical jet aircraft cruising altitudes of 35,000–40,000 ft.

The higher the radar's operating frequency (and the shorter its wavelength) is, then the poorer its all-weather performance becomes. This is illustrated by the fact that the intensity of radar echoes from rain showers increases proportionately to the fourth power as radar operating frequency is increased. However, higher frequency, shorter wavelength radars are required to provide the accuracy and resolution necessary to precisely track targets, guide missiles, or direct anti-aircraft gun systems. Thus radars associated with surface-to-air missile (SAM) systems (such as the Soviet Fan Song, Straight Flush, and Flap Lid) operate at much higher frequencies (1–10 GHz) and shorter wavelengths. Their detection ranges are typically under 100 n miles. Airborne interceptor radars operate at yet even higher frequencies (10–20 GHz).

Typically, the elements of an integrated air defense system are located so as to optimize the overall defense of high-priority target complexes. Individual radars are deployed to complement one another as they perform their roles of surveillance, acquisition, and fire control. A typical IADS layout is shown in Fig. A5.

By convention, electromagnetic frequencies are grouped into bands. Commonly used radar bands and their designations are listed in Fig. A6. A selection of Soviet-developed radar systems are also included as they show typical operating frequencies/bands for various types of radar applications. The ground-based surveillance radars can be seen to operate at the lower (VHF/UHF) frequencies (for longer detection ranges) with SAM and airborne radars operating at progressively higher frequencies (S, C, X, and Ku bands) for better tracking accuracy and higher lethality.

RADAR DETECTION RANGE

As noted in the preceding discussion, the maximum range ($R_{\max}$) at which a radar can detect a target is dependent on the radar's operating frequency/wavelength. However, there are a number of other radar and target characteristics that are also critical to defining a specific radar's detection range. These include the maximum power transmitted by the radar Pt (in watts); the radar antenna's gain G; the minimum level of received power that can be detected by the radar in the pres-

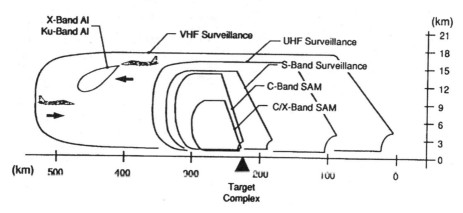

Fig. A5 **Threat radars are deployed to complement one another in an integrated air defense system.**

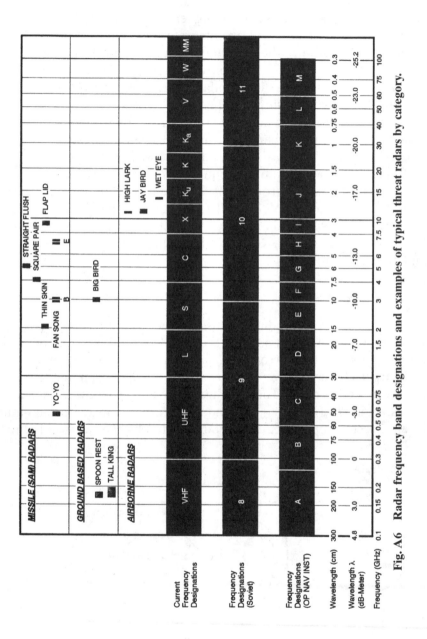

Fig. A6   Radar frequency band designations and examples of typical threat radars by category.

ence of noisc $Pr_{\min}$ (also in watts); and the electrical cross section of the radar's target. The latter is commonly known as the radar cross section (RCS) or the target's σ. It is usually expressed in square meters. Knowing these parameters, the maximum detection range can be determined from the well-known radar range equation, which, in a simple form, can be expressed as

$$R_{\max} = [Pt\ G^2\ \lambda^2\ \sigma\ /\ (4\ \pi)^3\ Pr_{\min}]^{\frac{1}{4}} \qquad (A2)$$

It can be observed that the maximum detection range of a radar varies as a function of the fourth root of the target's RCS. Thus achieving tactically significant reductions in detection ranges of threat radars requires huge (orders of magnitude) reductions in aircraft RCS. For instance, reducing the RCS of an aircraft by half will only reduce the detection range of a threat radar by about 16%.

ANTENNA GAIN

The effectiveness of a radar is dependent on the ability of its antenna to concentrate or focus electromagnetic energy in a narrow angular region. Antenna gain G is a measure of this ability.* For a circular antenna, typically used in fire control radars, the gain is dependent on both the antenna's physical size and its operating wavelength and can be approximated by the relationship in Eq. A3:

$$G \approx 5(D/\lambda)^2 \qquad (A3)$$

Increasing antenna size (where D is antenna diameter in meters) increases gain (and thus effectiveness). However, there are feasibility limits on the maximum size for tactical radar antennas, especially in airborne applications such as air intercept radars and missile installations. Thus, most fire control radars are designed to operate at shorter wavelengths (higher frequencies) to achieve necessary antenna gain and compensate for physical restrictions on radar antenna size. This also has the added advantage of providing the resolution and enhanced tracking accuracy needed for weapons guidance applications.

RADAR CROSS SECTION

There is only one factor in the radar range equation that is directly related to the specific characteristics of the radar's target. This factor is

---

* Efficient antennas can have gains of over 40,000 (40 dB) when compared to an omnidirectional antenna.

the RCS σ. Therefore, reducing the magnitude of RCS is a basic tenet of all radar signature reduction efforts. Because the RCS of actual radar targets encountered in day-to-day operations ranges from the very large (thousands of square meters for large aircraft and ships) to hundredths or even thousandths of a square meter (for birds and insects), RCS is often expressed in logarithmic form in decibels relative to one square meter or σ_{dBsm} as seen in Eq. A4:

$$\sigma_{dBsm} = 10 \log_{10} \sigma \qquad (A4)$$

Using this convention, the terminology commonly used to describe a radar target having an RCS of say 1000 m^2 (10^3 m^2) would be to say its RCS is 30 dBsm. Using the preceding, it can be shown that a 10-dB reduction in radar signature corresponds to a 90% reduction from the original target signature, 20 dB corresponds to a 99% reduction in target size, 30 dB is equivalent to a 99.9% reduction, and so forth.

The effect of reductions in target RCS on radar detection range can be determined using the following expression:

$$R_2/R_1 = (\sigma_2/\sigma_1)^{\frac{1}{4}} \qquad (A5)$$

where R_1 is the maximum range at which a target of RCS σ_1 can be detected in a given noise background. R_2 is the new maximum range at which a target aircraft can be detected if its RCS is reduced from σ_1 to σ_2.

Using the preceding expression, a one order of magnitude (10 dB) decrease in RCS ($\sigma_2/\sigma_1 = 0.1$) can be seen to reduce the capability of a threat radar to 56% of its previous detection range. A two order of magnitude (20 dB) decrease ($\sigma_2/\sigma_1 = 0.01$) reduces detection range to about 32% of the original capability. To reduce the detection range of a threat radar to 15% of its original capability necessitates a reduction in radar signature of about three orders of magnitude (30 dB); a fourth order reduction (40 dB) is needed if the detection range is to be reduced to 10% of the original capability. This relationship is graphically shown in Fig. A7.

To further put this in perspective, if one could reduce the RCS of an object having a RCS of 10 m^2 (similar to that of a typical fighter plane viewed from the nose) and if it were feasible to achieve a fourth order of magnitude (40 dB) signature reduction, the new object would have a radar signature approaching that of a very small bird or even of certain larger insects. The latter represents a reduction in the original radar signature on the order of 10,000 times. For reference, the Blue Winged Locust is about 20 mm long (slightly less than one inch). It has

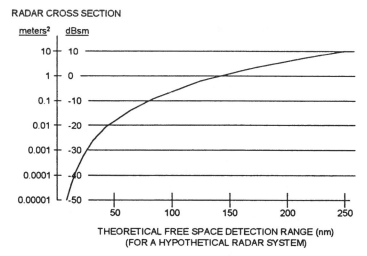

Fig. A7   Relative radar detection range as a function of RCS.

a measured RCS at a typical radar fire control frequency (9.4 GHz) of −30 dBsm to −40 dBsm depending on aspect angle. For comparison, the RCS of larger birds is on the order of 0.1 m² (−10 dBsm); smaller birds fall within the region of −20 to −30 dBsm.[3,4]

A vast RCS variation (covering a range of over 70 dB or seven orders of magnitude) exists between various groups of man-made and naturally occurring airborne objects as depicted in Table A1.

HAVE BLUE AND RADAR CROSS SECTION REDUCTION

Today, design RCS (and infrared) signature levels are routinely specified in weapon system development efforts; they are similar to other performance parameters in that they depend on the overall mission requirements of the specific weapon system. In general, the RCS requirement for a system does not require it to be totally invisible to all conceivable threat radars at all times (which is practically impossible) but instead to negate threat radars via a combination of techniques. These could include (in addition to RCS reduction) mission routing,* support jamming, and possibly the use of selective (intelligent) onboard jamming. The Have Blue technology demonstrator flight-test aircraft had

---

* An advanced automated mission planning system was also developed as an integral part of the F-117A system concept. It supported optimum employment of the inherent low RCS of the aircraft. The mission planning system accommodated rapidly updated radar threat system locations and operating parameters. It incorporated terrain data for areas of operational interest and used actual F-117A aircraft RCS patterns to optimize mission route profiles to best counter threat radar detection and engagement possibilities.

**Table A1   Variation in RCS between various
groups of natural and man-made objects**

RCS[a]		
m^2	dBsm	Object
100+	20+	
		Conventional aircraft
10	10	
1	0	Missiles/skydivers
0.1	−10	
0.01	−20	Birds/artillery shells
0.001	−30	
		Insects/bullets
0.0001	−40	
0.00001	−50	BBs

[a]Measured at 9.4 GHz.

no specific RCS goal or requirement; it was intended to demonstrate the lowest feasible aircraft radar signature that could be achieved while maintaining acceptable flying qualities. Have Blue would both validate the concept of low radar observability and generate valuable data for the follow-on F-117A strike aircraft and other stealth weapon systems.

Lockheed soon learned that a completely new outlook toward the entire aircraft design and weapon system development process was required to achieve very low radar signatures. In addition to basic aerodynamics, stability and control, aero-propulsion, weapons, and avionics integration considerations, it was vitally necessary to conceptually address the entire aircraft as a form of flying antenna. The task was to minimize radar reflections that could be received, detected, and tracked by threat radars while creating an aircraft capable of effectively accomplishing a demanding tactical mission. This approach affected all aspects of the design and demanded that intelligent tradeoffs and compromises be made in other areas such as aircraft performance and maintainability. The overall challenge facing the design team was to demonstrate the revolutionary new capability brought about through the use of stealth technologies while minimizing compromises and impacts on overall mission effectiveness.

To accomplish this daunting task, experts in the phenomena of electromagnetic scattering had to be fully integrated into the Have Blue and F-117A design teams. Electromagnetic scattering is a highly complex subject, yet it had to be fully understood to achieve truly meaningful RCS reductions.

RADAR SCATTERING PHENOMENA

Three phenomena (specular reflection, diffraction, and traveling and creeping waves) occur when the electromagnetic field generated by a radar encounters a target. The resultant scattering collectively accounts for the measured RCS of an aircraft.

Specular (mirror-like) reflections result when a radar wave is directly reflected from an object (Fig. A8). Diffraction occurs when electromagnetic energy encounters a sudden discontinuity or change in electrical impedance. Everyday examples of diffraction are rainbows, reflections from glass prisms, and the familiar phenomena that occurs when viewing an upright pencil in a glass of water: the image of the pencil seems to be broken at the air/water interface. Sources of sudden changes in electrical impedance that result in diffraction from an aircraft include wing leading and trailing edges, sharp corners on fins and tails, control surface gaps, and intersections between different airframe materials. An example of diffraction occurring at sharp edges is seen in Fig. A9.

Traveling and creeping waves are propagated when electrical currents (induced by incident electromagnetic energy) move across the conductive surface of an object. The creeping wave phenomena is depicted in Fig. A10. The directions in which creeping and traveling waves are reflected are dependent on a number of factors, including the geometrical shape of the object, its physical dimensions, and the angle of illumination and wavelength of the radar. An example of propagation of a traveling wave from a simple object is shown in Fig. A11.

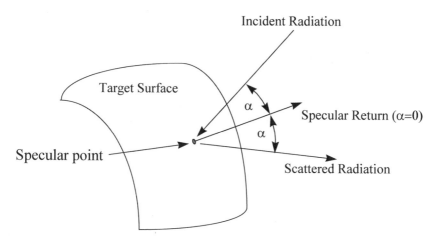

Fig. A8   Specular reflection.[5]

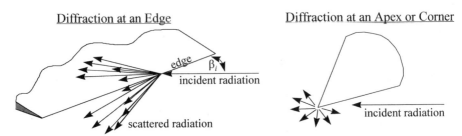

Fig. A9   Diffraction occurs at the intersections of sharp edges on an object.[5]

As can be seen from the figure, there is a relationship between the angle at which the traveling wave is propagated, the illuminating radar's wavelength, and the physical dimensions of the illuminated object. Obviously, reflection of traveling waves from more complex objects like aircraft present a much more difficult problem than this simple case.

AIRCRAFT RADAR CROSS SECTION

The RCS of an aircraft is the sum of all reflected radar energy from all scattering centers on the aircraft as depicted in Fig. A12. Many factors affect an aircraft's RCS, including its physical size and configuration (geometric shape). Specific design details such as engine inlet and nozzle geometry; the presence of canopies and other openings (such as windows and sensor ports); the location and orientation of hatches, control surfaces, maintenance access and weapons bay doors; the presence and location of external stores such as drop tanks and weapons; the location and types of radomes and antennas; and the

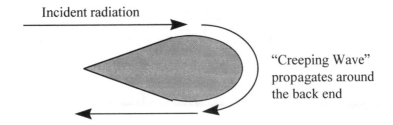

Fig. A10   Creeping waves are generated by induced surface currents on objects having rounded physical cross sections.

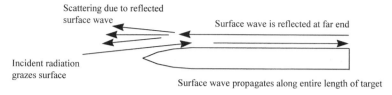

Fig. A11 Traveling waves are created by induced surface currents on a radar target.[5]

materials used in the aircraft's construction will also influence radar signature. Therefore all aspects of an aircraft's design down to the finest details (including even seemingly minor items such as navigation lights) must be assessed in the context of their contribution to overall RCS. Detail design of these individual components to ensure proper functionality as well as low RCS can represent a significant challenge. In the F-117A development, pitot tubes, inlet screens, and screens covering the infrared search and track system caused serious concerns.

The measured RCS of an actual aircraft is also dependent on its three-dimensional orientation relative to the threat radar as well as the threat radar's operating frequency and its polarization. Polarization describes the orientation of the electric field vector component of the radiated electromagnetic field; radar antennas are designed to radiate in specific orientations or polarization patterns. Polarization is commonly described as being either horizontal or vertical. Electromagnetic waves have electrical and magnetic field components whose vectors are orthogonal and in phase with one another; these, in turn, are perpendicular to the direction of propagation of the wave.[4,6]

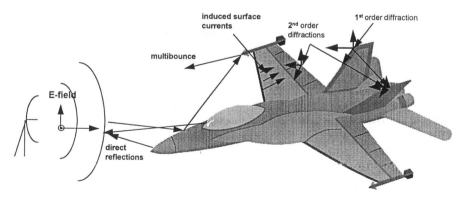

Fig. A12 The combination of various scattering phenomena produce an aircraft's RCS.[5]

Most modern radars can switch polarization between the horizontal and the vertical. This feature is often employed to improve target detection and tracking because target signature (RCS) is often greater in one polarization than in the other depending on the arrangement of the target's geometry and other physical features. Many radars also incorporate a feature known as circular polarization; this causes the emitted electromagnetic field to rotate. Circular polarization is either right- or left-handed depending on the sense of the rotation of the electromagnetic field[4,6]; it has long been employed to reduce the effects of extraneous noise caused by environmental conditions such as heavy rain or flocks of birds. Circular polarization can reduce echoes caused by rain by as much as 15 dB at C-band frequencies with only about a 3-dB loss in aircraft return. Typical birds produce radar signatures ranging from about −20 dBsm (duck) to −30 dBsm (starling); circular polarization reduces their return by about 3 dB (the same as it reduces aircraft returns).[3]

A variety of analytical prediction techniques using high-speed digital computers to determine the RCS of complex objects were developed, refined, and implemented in a variety of companies and government agencies during the 1970s. They would play a significant role in the design of Have Blue. RCS prediction capabilities have been continually improved since then; however, even the most refined of these techniques do not completely account for higher order edge diffraction effects or reflections due to surface and creeping waves; thus extensive RCS testing using high fidelity models is needed. Over the years since the early 1960s, a vast array of government and industry RCS measurement ranges (ranging from relatively miniature indoor facilities to vast outdoor complexes in remote secure areas) have been developed and built for this purpose.[5]

The earliest RCS prediction methods were based on optics analogies; however, these failed when the scattering direction moved too far from the specular direction. Methodologies to more accurately analyze diffraction effects at sharp edges were developed, based on pioneering work done by Ufimtsev (his "Physical Theory of Diffraction") in the Soviet Union and other researchers in the United States. Lockheed's ECHO RCS prediction model was used in the design of both Have Blue and the F-117A.[1] However, even refined techniques could not fully address diffraction at the intersections of sharp edged surfaces and other complex shapes such as curved inlets. A variety of more refined analytical prediction techniques have been developed since then to determine the RCS of more complex objects. However, even today, the most refined of these techniques do not completely account for higher order edge diffraction effects or reflections due to surface and creeping waves. Cavities (such as inlets and exhausts) and the myriad of smaller aircraft details (gaps, drain holes, etc.) present particular complications; therefore, extensive RCS testing, highly refined measurements, and careful analyses using highly detailed subscale and near/full-scale models are still essential even today.

MEASURED RADAR CROSS SECTION

Major changes in the magnitude of measured RCS occur when a complex object such as an aircraft varies its three-dimensional orientation relative to a radar. This phenomena is known as target scintillation. The scintillation in RCS pattern for the World War II–vintage Martin B-26 Marauder medium bomber is illustrated in polar format in Fig. A13. Measured RCS of a more modern aircraft, the British Hawker Siddeley H.S.125 corporate jet airliner, is shown in Fig. A14 in another commonly used RCS presentation (rectangular) format.

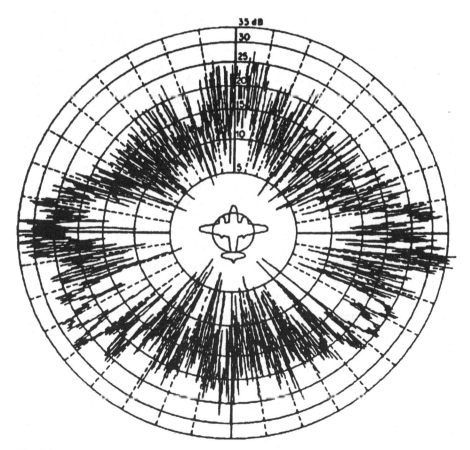

Fig. A13   RCS of the World War II Martin B-26 Marauder medium bomber measured at a frequency of 3 GHz (wavelength about 10 cm), vertical polarization.[4]

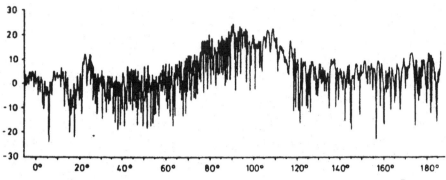

Fig. A14   H.S.125 RCS measured at 9.4 GHz, horizontal polarization.[7]

The RCS of the B-26 varies from about 20–25 dBsm (several hundred square meters) measured from the nose to 30–35 dBsm (well over a thousand square meters) on the beam. The H.S.125 radar signature ranges between about 3 m^2 head-on to over 20 m^2 on the beam. Conventional modern fighters and attack aircraft are generally similar in physical size to the B-26 and the H.S.125. These aircraft have radar cross sections that are similar to that of the H.S.125 when they are in clean configuration. External stores (e.g., drop tanks, bombs, missiles) can increase the signature of a conventional fighter to approximately 10 m^2 (10 dBsm) when viewed head-on and to over 100 m^2 (20 dBsm) when viewed on the beam.

Note that both radar frequency and polarization are specified with the RCS data for both aircraft. Knowledge of these parameters is essential if meaningful conclusions are to be drawn from RCS measurements because RCS varies with both frequency and polarization.

RADAR SIGNATURE REDUCTION TECHNIQUES

Over the years extensive research and development efforts in many countries had focused on reducing radar signatures; some of these efforts reached back as far as World War II. The German Type XXI U-boat, just entering Kriegsmarine operational service at the end of the war in early 1945, used so-called Wesch mats containing radar absorbent material to mask its snorkel head from the 9-cm wavelength (about 3.5 GHz) S-band search radars carried by allied anti-submarine patrol aircraft. RCS was reportedly reduced by 70%; this would have reduced the detection range of allied anti-submarine warfare (ASW) radars by about 25%. This may have had a significant impact on U-boat survivability because ASW aircraft operated at low altitudes where line of sight limitations and rough sea effects complicated radar detection probabilities. The

Type XXI U-boats were also fitted with retractable search radars and a radar warning receiver to detect emissions from allied search radars. The warning receiver was mounted on the snorkel and could detect radar signals between 1 and 4 GHz while the U-boat was submerged.[8]

These early attempts involved application of radar absorbent materials (RAMs) to the airframe or the use of radar absorbent structures (RASs). There are two types of RAM: magnetic and electrical. As their names imply, these function by absorbing either the magnetic or electric component of the electromagnetic wave. Early magnetic RAM consisted of iron particles embedded in a carbon matrix; it functioned by converting the incident radar energy into heat. One type of electric RAM is based on the use of myriad small electric dipoles embedded in a nonconductive matrix; when subjected to incident radar waves, the dipoles dissipate electromagnetic energy by generating twisting moments on the matrix material.[9]

The German Gotha 229 tailless flying wing jet fighter (which flew in prototype form as the Horten IX in January 1945) appears to be the first such attempt. Parasitic RAM was also tried on the Lockheed U-2.[10] However, to achieve truly low RCS, shaping of both the aircraft planform *and* its physical cross section was found to be necessary to redirect (forward scatter) incident radar energy away from threat radar. The conceptual value of shaping to reduce (or increase) RCS was well known; it was the basis for the simple "corner reflector," which had commonly been used for years to increase return from radar targets. To be truly effective, such shaping would have to be incorporated into the basic design.[9]

A third RCS reduction technique, use of an active cancellation avionics system to neutralize reflected radar energy, is extensively discussed in radar literature. Active cancellation systems are very complex; they require the development of a sophisticated onboard avionics system that senses the incident radar energy's direction of propagation, its frequency, its amplitude, and its polarization. This system must also know its own RCS as a function of these same parameters. The system then generates a return signal that cancels out its RCS. Active RCS cancellation systems had been studied and analyzed, but they had serious limitations at the time of F-117A development. These included inadequate computer storage capacity and system speed, issues with timing of the cancellation signal, and difficulties in dealing with multiple and/or simultaneous radar threats.[9]

In 1957, well before the 1960 Soviet shootdown of Gary Powers's U-2, the CIA asked Lockheed to conduct a low-observable U-2 replacement design study known as Project Gusto. This subsonic flying wing employed a combination of shaping and RAM; it was dropped when it was decided that a combination of supersonic speed and low RCS was required for the strategic reconnaissance overflight mission. This was the beginning of the Lockheed A-12/SR-71 development effort. Extensive airframe shaping and RAM were used by Lockheed on the A-12/SR-71 family and its D-21 supersonic reconnaissance drone (designed during the late 1950s through the early 1960s)[10]; however, radar signatures were not reduced to the point that true low observability was achieved. A different design approach was done for the Office of Naval Research by McDonnell in the early 1970s.[11] However, none of these efforts, which are illustrated in Fig. A15, succeeded in achieving the orders of magnitude reduction in aircraft RCS that were needed to ensure high survivability.

   In both Have Blue and the F-117A aircraft development, Lockheed focused its design efforts on the faceted shaping concept along with extensive application of parasitic magnetic RAM to reduce specular

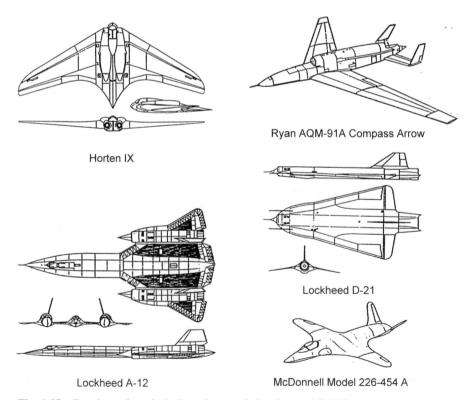

Horten IX

Ryan AQM-91A Compass Arrow

Lockheed D-21

Lockheed A-12

McDonnell Model 226-454 A

Fig. A15   Previous aircraft designs that used shaping and RAM to lower radar signatures.[10,11]

emissions and control surface currents generated by incident electro-magnetic energy. The Have Blue program conclusively proved that shaping, when properly implemented, has a powerful effect on RCS reduction. Shaping is, of course, also fundamental to the basic design of any aircraft; it determines overall aerodynamic performance and strongly influences basic stability and control characteristics. Thus, sophisticated tradeoffs are necessary during the aircraft design process to balance aerodynamic performance, aircraft handling qualities, and radar signature requirements.

Shaping to reduce RCS requires that both the planform of the aircraft *and* its physical cross section be properly configured to minimize radar reflections that could be detected by threat receivers. Figure A16, from the Lockheed patent application, shows a cross section of a Have Blue–type faceted aircraft that incorporates such shaping (this corresponds with item 4 in Fig. A1). It is particularly important that airframe features (such as the engine inlets and exhaust area) that are potentially large contributors to RCS be properly shaped. The Have Blue exhaust approach is seen in Fig. A17; it featured a flattened slot convergent nozzle having an aspect ratio of 17 to 1. This radically configured exhaust installation was also the primary means of reducing infrared signature (as discussed in the next section) and required some performance compromises during propulsion system design and integration. As previously noted, the engine inlets featured gridded covers treated with RAM to prevent incident radar waves from entering the inlets and being subsequently reflected.

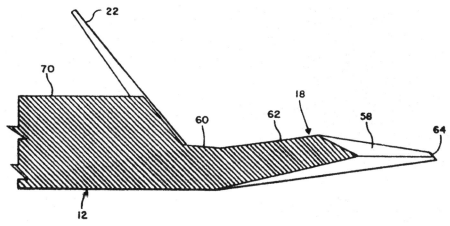

Fig. A16   Aircraft cross section showing application of shaping to reduce RCS.[2]

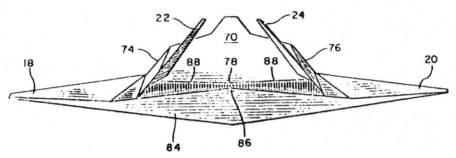

Fig. A17   Have Blue exhaust nozzle integration.[2]

INFRARED RADIATION

Like radio waves, infrared radiation is also part of the electromagnetic spectrum, with wavelengths between 0.77 to 1000 μ (equivalent to 10^{-6} m). Infrared waves are subdivided into near, mid, far, and extreme, as shown in Fig. A18, although the exact dividing points vary by convention.

Infrared radiation is emitted by all bodies at temperatures above absolute zero (−453°F or −273°C). The total amount and distribution of thermal radiation as a function of wavelength is affected by the material and the temperature of the body. For solid substances (continuum emitters), such as aircraft surfaces, radiation is evenly distributed over a wide wavelength band; for gases (line emitters), such as engine exhaust plumes, the radiation is emitted over a very small wavelength band.

The Stefan–Boltzmann law approximates emitted radiation energy M (W/m^2) by

$$M = \varepsilon \sigma T^4 \tag{A6}$$

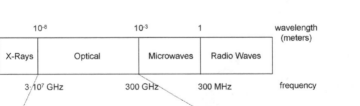

Fig. A18   Infrared spectrum.[12,13]

where the emissivity (a function of wavelength) is ε, the Stefan–Boltzmann constant is $\sigma = 5.67 \times 10^{-8}$ W/m^2-K^4, and temperature is on an absolute scale. Emissivity is a ratio of the radiation emitted by a surface to that emitted by a perfect radiating source at the same temperature. Thus, it is always less than unity.[13]

Although there are two contributors to the infrared signature of an aircraft, engine exhaust and aircraft skin heating due to atmospheric friction, at lower speeds the latter is almost negligible in comparison. The engine radiates infrared energy from its hot parts and from the exhaust plume itself (see Fig. A19). The fundamental variables available for reducing infrared radiation are temperature and emissivity, which is largely done by line of sight masking.

Air-to-air heat-seeking missiles typically are most sensitive in the middle of the infrared spectrum (3–5 μ), which is the natural frequency of hot carbon dioxide exhaust gases. Long wave infrared seekers (8–12 μ) detect the heat caused by solar heating and air friction on the aircraft skin. Both frequency bands are used on aircraft-mounted infrared seekers.[14]

Temperature reduction can produce a dramatic decrease in infrared energy. Equation (A6), however, is a great simplification because it neglects the frequency shift of radiation with temperature. At typical engine exhaust hot-metal temperatures, the exponential de-

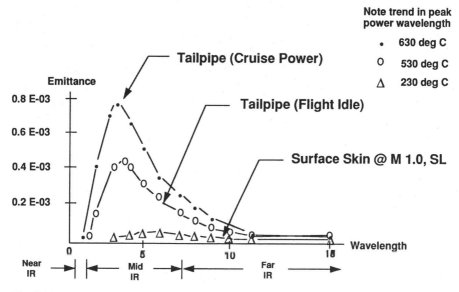

Fig. A19   Typical infrared radiation as a function of wavelength for a conventional aircraft.[12]

pendency will often be near eight rather than four. In any event, it is clear that a small reduction in temperature can have a much greater effect than the magnitude of emissivity reduction that can be reasonably affected. Today's state of the art in emissivity reduction within the desired frequency band can achieve emissivities on the order of 0.1 (Ref. 15).

For these reasons, attention in infrared radiation attenuation is primarily devoted to temperature reduction and masking. Reducing exhaust temperature is done in a variety of ways. Not using an augmentor, as with supercruise, significantly reduces the exhaust temperature. The more work that is pulled off by the turbine, such as with a turboprop or turboshaft engine, the cooler the exhaust is. For a jet engine where the hot exhaust is the propulsive force, this is less of an option. The Lockheed F-117A and the Northrop B-2 and YF-23 engines exhausted on the upper fuselage to prevent any hot parts being visible from below. Rectangular, or two-dimensional nozzles, also produce an unstable exhaust stream, which breaks up more rapidly, entraining ambient air for increased mixing and, therefore, more rapid cooling.[14]

While reducing the emissivity of a surface reduces the radiant energy, it also increases the amount of energy reflected from any hotter internal components. A careful optimization must therefore be made to determine the desired emissivity pattern inside the exhaust system. Short wavelength detectors are more effective at high temperatures, while long wavelengths are most effective at typical ambient atmospheric temperatures. This pattern must be such that it minimizes the signal within the typical detector frequency range as a function of both frequency and spatial location. In developing a coating that can attenuate the emissivity, the material must also not contribute to the radar return. Using a multilayer material for both radar and infrared reflection reduction requires a film thickness of a few angstroms.[15]

Engines, however, produce carbon deposits on the exhaust system surfaces, which have an extremely high emissivity. Therefore, engine exhaust would have to be reduced to essentially zero carbon particulate production, or the exhaust system would be covered with the high emissivity carbon coating after a few hours of operation. For this reason, emissivity attenuation is primarily attempted on the inlets and aircraft external surfaces. The aircraft skin friction and solar heating produce infrared radiation that is highly dependent on altitude and velocity. At the ATF supercruise point (approximately Mach 1.5), the aerodynamic heating can cause the skin temperature to rise by as much as 150°F, but this heating is largely negated by the cold (−70°F) temperatures at cruise altitude. At Mach 2, however, the aerodynamic heating can be 230°F; at Mach 2.5 it can be 500°F. Thus, skin heating can be highly significant at high airspeeds and is, in fact, the primary infrared contributor in the forward aspect sector when cruising without use of afterburners (see Fig. A20).[13–15]

In summary, infrared radiation can be reduced by careful application of temperature reduction design techniques and hot part component masking. As with RCS reduction, the dominant contributors must first be identified and then their emissions minimized through an integrated total system design approach.

ADVANCED TACTICAL FIGHTER CONCEPTS

Although the ATF prototypes were not themselves intended to be stealthy, they demonstrated the fundamental shaping principles that were representative of their intended production designs, which would have very low radar cross sections. Certain aspects of the Lockheed and McDonnell Douglas ATF design concepts represented somewhat differing approaches to future air combat. These different approaches (e.g., use of thrust vectoring) were made possible because the companies were allowed the flexibility to interpret the ATF system operational requirements.

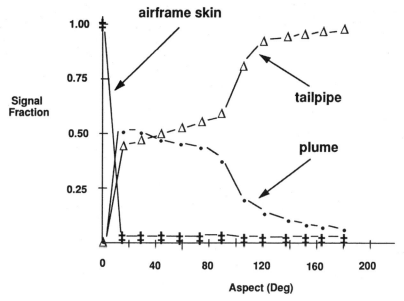

Fig. A20 Typical infrared return as a function of aspect angle for a conventional aircraft (0 deg corresponds to a nose-on view).[12]

Stealth, or signature reduction, was, however, to be achieved by the same physics: shaping and materials. Although radar absorbing materials and structures (RAMs and RASs) that would be part of the production aircraft were not used, the prototypes did demonstrate aggressive use of shaping for RCS reduction in their airframe designs.

Edge alignment is most readily apparent from looking at the aircraft. All angles were aligned to be parallel to the wing and tail angles, including the inlets, the exhaust nozzles, weapons bay doors, landing gear doors, etc. The canopy seam, bay doors, and other surface interfaces were sawtoothed to keep from having a surface that was perpendicular to the direction of flight. This surface alignment prevents any major radar return from being reflected back to a radar directly ahead of the aircraft. Instead it narrowly focuses nearly all reflected energy along a few angles.

Internal carriage of weapons was a major factor in reducing the radar cross section of the ATF aircraft. This ensured that the external airframe surfaces were smooth and thus prevented radar return from the ordnance and their ejection racks. The large wings would be able to carry enough fuel internally to roughly match the range of an F-15 equipped with external fuel tanks. A sharp chine that leads from the nose to the wings separated radar reflections from the top and bottom.

The engine face was hidden by long S-shaped ducts that curved inward and upward to minimize the radar return from the metal fan blades. The rectangular exhaust nozzles made the exhaust stream less stable, which entrained ambient air for more rapid mixing and, therefore, more rapid cooling. The canted vertical tails also helped to mask the hot exhaust gases from certain viewing angles.

The YF-23 (Fig. A21) was the less conventional appearing of the two aircraft, with a strikingly stealthy appearance. Its clean lines, prominent area ruled airframe, and minimum number of angled planes resulted in reduced drag and a radar signature that was narrowly focused along four distinct angles. Calculations and wind-tunnel and flight-testing demonstrated YF-23 maneuverability levels that were consistent with the intent of ATF requirements; however, flight at extremely high angles of attack was not considered an essential part of the Northrop design concept, and this aspect was consequently not flight demonstrated during the YF-23 Dem/Val effort.

The YF-23 incorporated a single weapons bay for the AIM-120 AMRAAM radar guided missiles. This was considered to be an adequate representation of the operational F-23 concept; the production version would have also had a second, smaller bay just forward of the main bay for the two AIM-9 Sidewinder infrared guided missiles.

The YF-23's engine inlets were under the forward inboard section of the wings. The engines exhausted through chevron-shaped noz-

Fig. A21   Planform view of the YF-23.

zles in a trough over a flat deck, similar to Northrop's B-2 exhausts, shielding the exhausts from infrared seekers from below as long as possible to maximize exhaust mixing with the atmospheric air. The serrated auxiliary inlets and excess air spill mechanisms were also developed to be low observable, but still function as efficiently as possible.

The YF-22 (Fig. A22), on the other hand, tried to balance stealth with maneuverability at very high angles of attack in order to be exceptionally lethal from well beyond visual range as well as in close-in dogfights with highly maneuverable opponents. The large side-mounted (or caret) inlets were located about even with the nose landing gear, a feature that could help to minimize foreign object damage to the engines from debris being thrown upward by the nose wheels. Again, as on the YF-23, the air inlet ducts curved inward and upward to the engine front faces, and serrated-edge auxiliary inlets and spill screens were designed to minimize overall radar signature. Nozzle designs for both the YF-22 and the YF-23 were chevron-shaped to minimize infrared and radar signatures.

Fig. A22   Overhead view of the YF-22.

The YF-22 surfaces were aligned so that the leading edges of the wing and tail had identical sweep angles, as did their trailing edges. This feature narrowly focused nearly all reflected radar energy from these surfaces along these four angles (and their opposites for a total of eight reflected radio frequency energy spikes).

The YF-22 incorporated three weapons bays: a bay for a single AIM-9 Sidewinder on the side of either inlet and a main bay for the AIM-120 AMRAAM missiles.

Conclusion

By the mid-1970s, modern radar-controlled integrated air defense systems were seriously threatening the capability of manned aircraft to accomplish vital combat missions. This was especially true for aircraft that were required to take the air battle into enemy territory to accomplish the role for which they had been designed. These roles were vital to the conduct of operations at the core of the Air Force's rationale for existence and included long range strike (attacks on strategic and deep interdiction targets), interdiction in support of the land battle (tactical interdiction), and offensive counterair operations (against both airborne and surface targets that threatened friendly air operations).

The Have Blue program demonstrated the feasibility of a very low observable combat aircraft that could render most known air defense radar guidance systems ineffective. It paved the way for the development of highly lethal and survivable operational combat aircraft (such as the F-117A, the B-2, and the F-22A). These aircraft ensure that vital Air Force roles can be accomplished well into the next century. They were designed using newly developed analytical capabilities pioneered in Have Blue. Airframe shaping approaches, along with extensive application of RAM and integration of radar absorbent structures, produced major reductions in radar signature (several orders of magnitude) compared to conventional aircraft. To achieve these very low signatures, weapons had to be carried internally, and all airframe features that could create undesirable reflections (including sensors and propulsion systems) had to be specially designed and appropriately treated. This was the challenge that faced the F-117A, and later the B-2 and F-22, design teams as they moved on to create the modern low-observable U.S. Air Force.

REFERENCES

[1]Aronstein, D. C., and Piccirillo, A. C., *Have Blue and the F-117A: Evolution of the "Stealth Fighter,"* AIAA, Reston, VA, 1997.

[2]Lockheed Corp., *Patent Application (Docket No. P-01-1532) for a "Vehicle" (Inventors: Richard Scherrer, Denys Overholster, and Kenneth E. Watson),* Burbank, CA, Jan. 16, 1979, filed with the commissioner of U.S. Patents and Trademark Office on Feb. 11, 1979.

[3]Blackwell, F., Houghton, E. W., and Wilmot, T. A., "Birdstrike and the Radar Properties of Birds," *International Conf. on Radar—Present and Future,* CP No. 106, Inst. of Electrical Engineers, London, 1973.

[4]Skolnik, M., *Radar Handbook,* 2nd ed., McGraw-Hill, New York, 1990.

[5]*RATSCAT Signature Measurement Short Course Notes,* 46th Test Group, U.S. Air Force, Holloman AFB, undated.

[6]Stimson, G. W., *Introduction to Airborne Radar,* Hughes Aircraft Co., El Segundo, CA, 1983.

[7]Cram, L. A., Johnson, R. H., and Woolcock, S. C., "Radio Scale Modelling in Support of Radar System Design and Assessment of Performance," *Inter-*

national Conf. on Radar Present and Future, CP No. 106, Inst. of Electrical Engineers, London, 1973.

[8]Köhl, F., and Rössler, E., The Type XXI U-Boat, U.S. Naval Inst. Press, Annapolis, MD, 1991.

[9]Knott, E. F., Shaeffer, J. F., and Tuley, M. T., Radar Cross Section—Its Prediction, Measurement and Reduction, Artech House, Inc., Norwood, MA, 1985.

[10]Miller, J., Lockheed's Skunk Works: The First Fifty Years, Aerofax, Inc., Arlington, TX, 1993.

[11]Woods, J., and Rose, R. G., Quiet Attack Aircraft Radar Cross Section Test Results, McDonnell Douglas Corp., Technical Rept. MDC A2949-2, St. Louis, MO, Nov. 15, 1974.

[12]Nelson, T., "A General Overview of Infrared Applications," ANSER, Inc., Arlington, VA, 1993.

[13]Ball, R. E., The Fundamentals of Combat Aircraft Survivability Analysis and Design, AIAA Education Series, AIAA, New York, 1985.

[14]Sweetman, B., YF-22 and YF-23 Advanced Tactical Fighters: Stealth, Speed and Agility for Air Superiority, Motorbooks International, Osceola, WI, 1991.

[15]Brown, A., "Fundamentals of Stealth Design," Lockheed Corp., Burbank, CA, 1992.

BIBLIOGRAPHY

Abrams, R. (Director of Flight Test, Lockheed Advanced Development Co.), *YF-22A Prototype Advanced Tactical Fighter Demonstration/Validation Flight Test Program Overview,* Lockheed Corp., Palmdale, CA, 1991.

Abrams, R., and Miller, J., *Lockheed (General Dynamics/Boeing) F-22,* Aerofax Extra, No. 5, Aerofax-Midland Publishing, Ltd., Leicester, UK, 1992.

Adams, A., "Advanced Technology Engine Studies—Final Review," Pratt and Whitney, West Palm Beach, FL, Feb. 23, 1982. (P&W Proprietary; all information used has been approved for public release.)

Adams, J. V., Col. USAF (TAC/DRD), to TAC/CC, /CV, /CS, /XP, Staff Summary Sheet, Subject: *Future Fighter Alternatives Study Steering Committee Meeting,* Nov. 3, 1980.

"Advanced Fighters: Big Plum for Aerospace," *High Technology,* April 1984, p. 49.

"Air Force Plans to Limit Tactical Fighter Costs," *Aviation Week and Space Technology,* Vol. 123, No. 4, July 29, 1985, p. 14.

Air Force Regulation AFR 57-1, *Statement of Operational Need (SON),* HQ USAF, June 12, 1979.

Air Force Requests Proposals for Next Generation Fighter, ATF press release, number illegible, Oct. 8, 1985.

"Air Force Wants More INEWS Teaming," *Aerospace Daily,* Vol. 138, No. 62, June 26, 1986, p. 489.

Air-to-Surface (ATS) Technology Evaluation and Integration Study, Final Report for Period June 14, 1976–Oct. 14, 1977, Grumman Aerospace Corp., Rept. No. AFFDL-TR-77-131, Bethpage, NY, Jan. 15, 1978.

Air-to-Surface (ATS) Technology Evaluation and Integration Study, Presentation Material, McDonnell Douglas Corp., Rept. No. MDC A4285-2, St. Louis, MO, Jan. 27, 1977.

Air-to-Surface Technology Integration and Evaluation Study, Third Quarter Status Report, General Dynamics, Rept. No. FZM-6667, Fort Worth, TX, May 11, 1977.

"Annex B—Program Life-Cycle Cost Estimate Summary ($M)," (Congressional C-Document), Office of the Assistant Secretary of the Air Force (Acquisition), Fighter Div. (SAF/AQPF), Pentagon, Washington, DC, Aug. 1994.

Aronstein, D. C., *Development of Radar Cross Section Prediction Methodology,* ANSER, Inc., Arlington, VA, June 1996.

Aronstein, D. C., and Piccirillo, A. C., *Comments and Observations on Flight Demonstrator Programs,* ANSER Rept. prepared for the Joint Advanced Strike Technology Program Office, Arlington, VA, Oct. 25, 1994.

Aronstein, D. C., and Piccirillo, A. C., *Have Blue and the F-117A: Evolution of the "Stealth Fighter,"* AIAA, Reston, VA, 1997.

Aronstein, D. C., and Piccirillo, A. C., *The Lightweight Fighter Program: A Successful Approach to Fighter Technology Transition*, AIAA, Reston, VA, 1997.

"ASD Awards Two Contracts for Demonstrator Engines," *Skywrighter,* Vol. 24, No. 40, Oct. 7, 1983, p. 1.

"ATF Is Delayed to Add Stealth Technologies," *Defense Week,* June 27, 1983, pp. 4, 5.

"ATF Prototypes Burn One-Third Less Fuel in Supercruise," *Aviation Week and Space Technology,* Vol. 133, No. 83, Dec. 10, 1990.

Ball, R. E., *The Fundamentals of Combat Aircraft Survivability Analysis and Design,* AIAA Education Series, AIAA, New York, 1985.

Banvard, K., "McClellan Will Maintain New Generation of Fighters," *Sacramento Union,* May 5, 1984, pp. A1, A2.

Bazley, Gen. USAF, Donnelly, Gen. USAF, and Kelley, Lt. Gen. USAF, Memorandum to Gen. Gabriel, May 14, 1985. SECRET. (Unclassified information only used from this source.)

Bedard, P., "Industry Raps ATF Plans," *Defense Week,* Vol. 7, No. 43, Nov. 3, 1986, p. 3.

Blackwell, F., Houghton, E. W., and Wilmot, T. A., "Birdstrike and the Radar Properties of Birds," *International Conf. on Radar—Present and Future,* CP No. 106, Inst. of Electrical Engineers, London, 1973.

Bond, D. F., "Risk, Cost Sway Airframe, Engine Choices for ATF," *Aviation Week and Space Technology,* Vol. 134, No. 17, April 29, 1991, pp. 20–21.

Borky, M. J., Col. USAF, *Advanced Tactical Fighter Demonstration/Validation Phase Avionics Program,* Press Conf. Briefing, Aug. 23, 1990.

Borky, M. J., Col. USAF, *Press Release on Avionics for the Advanced Tactical Fighter (ATF),* ASD 90-2347, Aug. 23, 1990.

"Breaking Barriers to Build the Next Air Force Fighter," *Business Week,* No. 2828, Feb. 13, 1984, p. 128.

Brown, A., *Fundamentals of Stealth Design,* Lockheed Corp., Burbank, CA, 1992.

Brown, D. A., "General Dynamics Evaluates Concepts for F-16 Successor," *Aviation Week and Space Technology,* Vol. 132, No. 24, June 11, 1990, pp. 21, 22.

Bruner, E. F., *Defense Department Bottom-Up Review: Results and Issues,* Congressional Research Service (CRS) Report for Congress, 93-839-F, Sept. 23, 1993.

Canan, J. W., "Toward the Totally Integrated Airplane," *Air Force Magazine,* Jan. 1984, pp. 34–41.

"CBO Challenges ATF Assumptions," *Aerospace Daily,* Vol. 132, No. 43, April 30, 1985, p. 340.

Charter for the Joint Advanced Strike Technology (JAST) Program, Approved by John M. Deutch (Deputy Secretary of Defense), Sheila E. Widnall (Secretary of the Air Force), and John H. Dalton (Secretary of the Navy), 1994.

Concept of Operation for the Advanced Tactical Fighter (ATF), U.S. Air Force Tactical Air Command, Deputy Chief of Staff (Plans), Jan. 1971. SECRET. (Unclassified information only used from this source.)

"Contractors Struggling," *Armed Forces Journal International,* Feb. 1989, p. 29.

Conver, S. K., Deputy Assistant Secretary of the Air Force (Programs and

Budget), Memorandum to Mr. Aldridge, Subject: *ATF Reliability, Maintainability, and Producibility,* Nov. 15, 1985.

Cooper, T. E., Assistant Secretary of the Air Force (Research, Development and Logistics), Memorandum to AF/RDCS, Subject: *Final Acquisition Action Approval for the Acquisition Plan for the Advanced Tactical Fighter (ATF) System,* Sept. 27, 1985.

Cooper, T. E., Assistant Secretary of the Air Force (Research, Development and Logistics), Memorandum for Vice Chief of Staff, Subject: *ATF AFSARC— Action Memorandum,* May 27, 1986.

Cram, L. A., Johnson, R. H., and Woolcock, S. C., "Radio Scale Modelling in Support of Radar System Design and Assessment of Performance," *International Conf. on Radar—Present and Future,* CP No. 106, Inst. of Electrical Engineers, London, 1973.

Department of Defense Major Aircraft Review, Hearings on National Defense Authorization Act for FY91, before the House Armed Services Committee, 101st Congress, 2nd Session, April 26, 1990, pp. 692, 693.

"DOD Will Accelerate INEWS program to Meet New ATF Prototype Schedule," *Inside the Pentagon,* Aug. 15, 1986, pp. 5, 6.

Dornheim, M. A., "Lockheed Team Will Test ATF Cockpit in Boeing 757 Flying Laboratory," *Aviation Week and Space Technology,* Vol. 125, No. 19, Nov. 10, 1986, pp. 22, 23.

Eaglet, R., Col. to HQ AFSC, HQ USAF, et al., Subject: *Future Fighter Alternatives Study,* Nov. 3, 1981.

"F 22 EMD Budget Timeline," Briefing Chart, Office of the Assistant Secretary of the Air Force (Acquisition), Fighter Div. (SAF/AQPF), Nov. 19, 1996.

"F-22 EMD Funding History," Briefing Chart, Office of the Assistant Secretary of the Air Force (Acquisition), Fighter Div. (SAF/AQPF), Nov. 19, 1996.

"F-22 Program Adjusted for the End of the Cold War," Briefing Chart, Office of the Assistant Secretary of the Air Force (Acquisition), Fighter Div. (SAF/AQPF), Nov. 19, 1996.

"F119 Configuration Reflects Balanced Design, Lessons From F100 Program," *Aviation Week and Space Technology,* Nov. 18, 1991.

Fact Sheet: Advanced Tactical Fighter, U.S. Air Force Systems Command, Aeronautical Systems Div., Office of Public Affairs (ASD/PA), PAM #84-011, Aug. 1984.

Ferguson, P. C., *Advanced Tactical Fighter/F-22 Annotated Chronology* (Draft), Office of History, Aeronautical Systems Center, Wright–Patterson AFB, OH, Aug. 1996.

Ferguson, P. C., From *Advanced Tactical Fighter (ATF) to F-22, Part I—To Milestone 0 and Beyond: 1970–1982,* Office of History, Aeronautical Systems Center, Wright–Patterson AFB, OH, May 1996.

Ferguson, P. C., *Oral History Interview: Brigadier General Claude M Bolton, Jr.,* Office of History, Aeronautical Systems Center, Wright–Patterson AFB, OH, Aug. 13, 1996.

"Fighter Weapons," *Aviation Week and Space Technology,* Letter to the Editor (author's name withheld), Vol. 122, No. 4, Jan. 28, 1985, p. 108.

"First YF-23 Completes Dem/Val, Demonstrates 'Surge' Capability," *Aviation Week and Space Technology*, Vol. 133, No. 24, Dec. 10, 1990, p. 83.

Flynn, J., and Eismeier, M., "Advanced Tactical Fighter Engine Development Program Comments on Proposed ANSER Paper," General Electric, Evendale, OH, June 13, 1997.

"From the Newly Released Bottom Up Review Report Section V: Modernization of Theatre Air Forces," *Defense Week*, Vol. 14, No. 41, Oct. 18, 1993.

General Accounting Office (GAO), *Aircraft Development: Navy's Participation in Air Force's Advanced Tactical Fighter Program*, GAO/NSIAD-90-54, GAO, Washington, DC, March 1990.

General Accounting Office (GAO), *Defense Acquisition Programs: Status of Selected Programs*, GAO/NSIAD-90-159, GAO, Washington, DC, June 1990.

General Accounting Office (GAO), *DOD Acquisition: Case Study of the Air Force Advanced Fighter Engine Program*, GAO/NSIAD-86-45S-13, GAO, Washington, DC, Aug. 25, 1986.

General Accounting Office (GAO), *Reasons for Recent Cost Growth in the Advanced Tactical Fighter Program*, GAO/NSIAD-91-138, GAO, Washington, DC, Feb. 1991.

General Electric, "PPSIP Master Plan," June 30, 1990. (GE Proprietary; all information used has been approved for public release.)

General Electric, "Propulsion Assessment for Tactical Systems, Final Report, Vol. 1, Program Summary," Sept. 1986. (GE Proprietary; all information used has been approved for public release.)

General Electric, "YF120 Engine Familiarization Workshop" Brief, June 12, 1996. (GE Proprietary; all information used has been approved for public release.)

"Getting INEWS Off the Ground," *Defense Electronics*, Vol. 18, No. 10, Oct. 1986, pp. 46–51.

Gideon, F. C., Lt. Col. USAF (Chief, Aircraft Div., DCS/Plans and Programs), Memorandum for the Record, *AFSC Council Minutes—New Fighter Aircraft—The Advanced Tactical Fighter Program*, Nov. 5, 1981.

Glasgow, E., "YF-22 Supermaneuverability," Draft Paper, Lockheed Corp., Undated.

Graves, J. T., Deputy Director of F-22 SPO, "Acquisition Executive Monthly Report (AEMR) for the F-22 (ATF) Program," Sept. 30, 1991.

Green, W., *New Observer's Book of Aircraft*, 1985 ed., Frederick Warne and Co., 1985.

"Growing ATF Program Transferred to Tactical Systems," *Air Force Times*, Vol. 45, No. 2, Jan. 1985.

Hardison, Deputy Undersecretary of Defense (Tactical Warfare Programs), to the Assistant Secretary of the Air Force (Research, Development and Logistics), Memorandum: *Mission Element Need Statement (MENS) for a New Fighter Aircraft*, Jan. 11, 1982.

Harris, R., "History of ATF/AX Engine Development Programs," NAVAIR document AIR-53511B/23421, Jan. 6, 1993.

Haystead, J., "Sleek, Swift, Superior: A Look at Tomorrow's Advanced Tactical Fighter for the USAF," *Journal of Electronic Defense*, Vol. 6, No. 11, Nov. 1983, pp. 53, 54.

Henry, D. D., Lt. Col. USAF (HQ USAF/RD&A), "Advanced Tactical Fighter Program," *Tactical Aircraft Research and Technology,* NASA Langley Research Center, NASA-CP-2164, Hampton, VA, March 1981. SECRET. (Unclassified information only used from this source.)

"Historical Fighter Production Cost Trend," Briefing Chart, U.S. Air Force Aeronautical Systems Div., Wright–Patterson AFB, OH, Undated.

Jane's Information Group, *Jane's All The World's Aircraft,* London, 1977–78, 1983–84, 1984–85, 1986–87, and 1996–97 eds.

Jasik, H., *Antenna Engineering Handbook,* McGraw-Hill, New York, 1961.

Kandebo, S. W., "Pratt's ATF Engine Victory Could Yield 1,500 Powerplants," *Aviation Week and Space Technology,* Vol. 134, No. 17, April 29, 1991, pp. 24–25.

Knott, E. F., Shaeffer, J. F., and Tuley, M. T., *Radar Cross Section—Its Prediction, Measurement and Reduction,* Artech House, Inc., Norwood, MA, 1985.

Köhl, F., and Rössler, E., *The Type XXI U-Boat,* The U.S. Naval Inst. Press, Annapolis, MD, 1991.

La Flam, D. J., Gupta, S. K., Paquette, R. G., and Raj, A., "Application of Stealth Technologies to High-Speed Patrol Craft," *EW Reference and Source Guide, A Supplement to the January 1996 Journal of Electronic Defense,* Norwood, MA, 1996.

Lenorovitz, J. M., "Enhanced Fighter," *Aviation Week and Space Technology,* Vol. 110, No. 15, April 9, 1979, p. 18.

Leuthauser, J., Maj. USAF/XRLA, *Talking Paper on Air Staff Proposals on ATF,* HQ USAF, Washington, DC, Dec. 5, 1985.

Lockheed Corp., *Patent Application (Docket No. P-01-1532) for a "Vehicle" (Inventors Richard Scherrer, Denys Overholtser, and Kenneth E. Watson),* Burbank, CA, Jan. 16, 1979, filed with the commissioner of U.S. Patents and Trademark Office on Feb. 11, 1979.

Lockheed Martin, "F-22 Chronology (including F119 engine dates)," Lockheed Martin F-22 Web Site, <http://www.lmasc. lmco.com/f22/>, April 1997.

"Lockheed's Kelly Johnson Building 'Stealth' Aircraft," *Aerospace Daily,* Vol. 80, No. 16, July 23, 1976, pp. 121, 122.

Lowndes, J. C., "Defense Studies Specialized Aircraft," *Aviation Week and Space Technology,* Vol. 115, No. 14, Oct. 5, 1981, pp. 81, 84, 85.

Lyons, R. F., Maj. USAF, *The Search for an Advanced Fighter: A History from the XF-108 to the Advanced Tactical Fighter,* Air Command and Staff College, Air Univ., Maxwell AFB, AL, April 1986.

McMullen, T., Lt. Gen. USAF, ASD Commander, to ATF Dem/Val Offerors: *Advanced Tactical Fighter (ATF) Program Restructure,* May 28, 1986.

Memorandum of Agreement Between Air Force Wright Aeronautical Laboratories and Aeronautical Systems Division's Deputy for Development Planning, Annex 2: Advanced Tactical Fighter (ATF) Program, Sept. 1981

Message, HQ TAC to USAF/RDQ, Subject: *ATF Program Management Directive New Addition,* April 22, 1982. SECRET. (Unclassified information only used from this source.)

Message, Vice Commander Air Force Systems Command (AFSC/CV) to

Commander Aeronautical Systems Div. (ASD/CC), Subject: *Fighter Alternatives Study,* Aug. 14, 1980.

Meyer, D. G., "Futuristic Multirole Aircraft Could Guard High Frontier," *Armed Forces Journal International,*Vol. 121, No. 2, Sept. 1983, p. 48.

Miller, B., "Advanced Tactical Fighter Studies Set," *Aviation Week and Space Technology,* Vol. 104, No. 3, Jan. 19, 1976, pp. 16, 17.

Miller, B., "USAF Plans Advanced Fighter Program," *Aviation Week and Space Technology,* Vol. 99, No. 3, July 16, 1973, pp. 14–16.

Miller, J., *Lockheed's Skunk Works: The First Fifty Years,* Aerofax, Inc., Arlington, TX, 1993.

Minutes of the Air Force F-22 Issues Conference, held at ANSER, Arlington, VA, July 14, 1994.

Mission Element Need Statement (MENS): Advanced Tactical Fighter, USAF/RDQ, Undated (circa early 1980).

Mission Element Need Statement (MENS) for New Fighter Aircraft, USAF/RDQ, first released July 1981. (The same MENS has appeared with cover memos bearing various dates in Sept. and Nov. 1981.)

Moorhouse, D. J., *Lessons Learned from the STOL and Maneuver Technology Demonstrator,* Flight Dynamics Directorate, Wright Lab., Rept. No. WL-TR-92-3027, Wright–Patterson AFB, OH, June 1993.

Moxon, J., "Goals Set for ATF Engine," *Flight International,* Vol. 125, No. 3910, April 14, 1984, pp. 1017, 1018.

Moxon, J., and Warwick, G., "Advanced Tactical Fighter," *Flight International,* Vol. 126, No. 3930, Oct. 20, 1984, pp. 1048–1061.

Mullin, S. N., "The Evolution of the F-22 Advanced Tactical Fighter," Wright Brothers Lecture, AIAA Paper 92-4188, Washington, DC, Aug. 24, 1992.

Mullin, S. N., Letter to D. C. Aronstein, Aug. 17, 1996.

Nelson, T., *A General Overview of Infrared Applications,* ANSER, Arlington, VA, 1993.

"New Advanced Fighter Blueprint Throws Key Schedules out of Kilter," *Defense News,* Vol. 1, No. 23, June 23, 1986, pp. 3, 4.

"New Soviet Aircraft Focus U.S. Attention on Advanced Fighter," *Aviation Week and Space Technology,* Vol. 120, No. 12, March 19, 1984, pp. 46, 47.

Nordmeyer, Capt. USAF, TAC/DRFG, to TAC/CC, /CV, and /CS, Staff Summary Sheet, Subject: *Advanced Tactical Fighters (ATF),* May 17, 1982.

Operational Requirements Document (ORD) (U) CAF 304-83-I/II-A (Rev. 2), *F-22 Advanced Tactical Fighter,* HQ ACC/DR SMO-22, Aug. 26, 1996, with Attachment: *Requirements Correlation Matrix.* SECRET. (Unclassified information only used from this source).

Patterson, R., Capt. USAF, *Wright Laboratory's Role/History in the Advanced Tactical Fighter Technology Development,* Wright Lab. Rept. No. WRDC-TM-90-603-TXT, Wright–Patterson AFB, OH, Jan. 1990.

Petty, J. S., Hill, R. J., Piccirillo, A. C., Col. USAF, and Fanning, A. E., Maj. USAF, "The Next Hot Fighter Engine," *Aerospace America,* Vol. 24, No. 6, June 1986.

Piccirillo, A. C., Col. USAF, "The Advanced Tactical Fighter: Design Goals and Technical Challenges," *Aerospace America,* Vol. 22, No. 11, Nov. 1984.

Piccirillo, A. C., Col. USAF, Memorandum for Mr. Tremaine, Subject: *ATF Status Report,* July 28, 1983.

Piccirillo, A. C., Col. USAF, USAF Aeronautical Systems Div. Advanced Tactical Fighter Program Office Weekly Activity Reports and Personal Notes, June 1983–Dec. 1986.

Plourde, G., "Advanced Tactical Fighter RFI Response," Pratt and Whitney, West Palm Beach, FL, Feb. 1982. (P&W Proprietary; all information used has been approved for public release.)

Pratt and Whitney, "Advanced Tactical Engine Studies (ATES), Final Report," Pratt and Whitney, West Palm Beach, FL, Sept. 10, 1982. (P&W Proprietary; all information used has been approved for public release.)

Pratt and Whitney, "The Core of Navy Air Power," Pratt and Whitney, West Palm Beach, FL, June 1, 1978. (P&W Proprietary; all information used has been approved for public release.)

Pratt and Whitney, "F119 Advanced Tactical Fighter Engine Program," Program Schedule, Pratt and Whitney, West Palm Beach, FL, Jan. 15, 1991.

Pratt and Whitney, "F119 Experimental Demo/Proto Engine Run Time History," Pratt and Whitney, West Palm Beach, FL, Jan. 1994. (P&W Proprietary; all information used has been approved for public release.)

Pratt and Whitney, "Proposal for Advanced Tactical Fighter Engine/Navy Advanced Tactical Fighter Engine, F119 Advanced Tactical Fighter Engine, Executive Summary, Vol. I," Pratt and Whitney, West Palm Beach, FL, Jan. 2, 1991. (P&W Proprietary; all information used has been approved for public release.)

Preliminary System Operational Concept (PSOC) for the Advanced Tactical Fighter, HQ Tactical Air Command, Deputy Chief of Staff (Plans), Rept. No. XPJ-C-85-001, Feb. 1, 1985. SECRET. (Unclassified information only used from this source.)

President's Blue Ribbon Commission on Defense Management, "An Interim Report to the President," Feb. 28, 1986.

Price, A., *Instruments of Darkness—The History of Electronic Warfare,* Macdonald and Jane's, London, 1978.

Program Management Directive No. R-Q R-Q 1057(1)/63242F, *Combat Aircraft Prototype (CAP),* USAF/RDQT, April 14, 1981.

Program Management Directive No. R-Q 7036(5)/63230F, *Advanced Tactical Fighter (ATF) Technologies,* USAF/RDQT, Aug. 24, 1982.

Raimundo, J., "McClellan Will Fix, Maintain Stealth Jets," *Sacramento Bee,* May 5, 1984, pp. A1, A14.

Rait, S., "Streamlining the Advanced Tactical Fighter," *Program Manager,* March–April 1985, pp. 2–5.

RATSCAT Signature Measurement Short Course Notes, 46th Test Group, U.S. Air Force, Holloman AFB, NM, 1993.

Rawles, J. W., "ATF Program Gathers Momentum," *Defense Electronics,* Vol. 20, No. 8, July 1988, pp. 43–51.

Read-Ahead Blue Book for the Defense Acquisition Board Advanced Tactical Fighter Milestone II Review, June 25, 1991.

Reis, R. J., Contracting Officer, HQ AFSC, to Industry Addressees, Subject: *Request for Information for Advanced Tactical Fighter Mission Analysis (RFI-ATF/MA-001),* May 21, 1981, with attached RFI Definition.

Robinson, C. A., "Air Force to Develop New Fighter," *Aviation Week and Space Technology,* March 30, 1981, pp. 1619.

Robinson, C. A., "USAF Reviews Progress of New Fighter," *Aviation Week and Space Technology,* Vol. 119, No. 22, Nov. 28, 1983, pp. 44–46, 51.

Scarborough, R., "AF Freezes ATF Designs," *Defense Week,* Vol. 8, No. 46, Nov. 23, 1987.

Schultz, J. B., "New Air Force Fighter to Employ Stealth Design and Advanced Electronics," *Defense Electronics,* Vol. 16, No. 4, April 1984, pp. 97–102.

Scott, W. B., "ATF Contractor Teams Accelerate Evaluations with Four Prototypes," *Aviation Week and Space Technology,* Vol. 133, No. 19, Nov. 5, 1990, p. 98.

Scott, W. B., "YF-23 Previews Design Feature of Future Fighters," *Aviation Week and Space Technology,* Vol. 133, No. 1, July 2, 1990, pp. 16–21.

Scott, W. B., "YF-23 Prototype Undergoes Flutter Tests, May Finish Flight Evaluations Next Month," *Aviation Week and Space Technology,* Vol. 133, No. 17, Oct. 22, 1990, pp. 114–115.

Semi-Annual History, AFSC Aeronautical Systems Div., Directorate of General Purpose and Airlift Systems Planning (ASD/XRL), Jan.–June 1972, p. 15.

Semi-Annual History, AFSC Aeronautical Systems Div., Directorate of Mission Analysis (ASD/XRM), July–Dec. 1982, p. 15.

Skantze, L. A., Gen. USAF, Letter to the Honorable Verne Orr, Secretary of the Air Force, May 10, 1985.

Skolnik, M., *Radar Handbook,* 2nd ed., McGraw-Hill, New York, 1990.

Smith, P., Brig. Gen. USAF, Deputy Director of Plans, DCS/P&O, to AF/RDQ, Subject: *For Comment MENS—AASF and ATSF,* May 20, 1981.

Snodgrass, M. A., Maj. USAF, *Information Paper: F-22 Weight Reduction Efforts,* SAF/AQPF, Oct. 22, 1993.

Stimson, G. W., *Introduction to Airborne Radar,* Hughes Aircraft Co., El Segundo, CA, 1983.

Sudheimer, R. H. (ASD), "Advanced Tactical Attack System Mission Analysis (ATASMA)," *Tactical Aircraft Research and Technology,* NASA Langley Research Center, NASA-CP-2164, Hampton, VA, March 1981. SECRET. (Unclassified information only used from this source.)

Sweetman, B., *YF-22 and YF-23 Advanced Tactical Fighters: Stealth, Speed and Agility for Air Superiority,* Motorbooks International, Osceola, WI, 1991.

Tremaine, S. A., ASD Deputy for Development Planning, to Industry Participants, Subject: *Request for Information (RFI) for Advanced Tactical Fighter (ATF) Mission Analysis,* May 21, 1981.

Turk, P., "Power for the ATF: GE and P&W Prepare for Battle," *Interavia,* Sept. 1984, pp. 895, 896.

Twigg, J. L., *To Fly and Fight: Norms, Institutions, and Fighter Aircraft Procurement in the United States, Russia and Japan,* Ph.D. Dissertation, Dept. of Political Science, Massachusetts Inst. of Technology, Cambridge, MA, Sept. 1994.

Ulsamer, E., "Scoping the Technology Baseline," *Air Force Magazine,* Vol. 65, No. 12, Dec. 1982, pp. 24, 27.

U.S. Air Force Press Release, "Integrated Electronic Warfare System (INEWS)," June 27, 1986.

"USAF Seeks Return to Fighter Traditions," *Flight International,* Vol. 124, No. 3883, Oct. 8, 1983, pp. 938–939.

Vartabedian, R., "Fierce Battle Waged over New Fighter," *Los Angeles Times,* April 15, 1984.

Wallace, Lt. Col. (TAC/DRFG) to TAC/CV, Staff Summary Sheet, Subject: *Future Fighter Alternatives Study,* Sept. 18, 1980.

Walley, R., Maj. USAF, *Information Paper: History of F-22 Unit Cost Estimates,* Office of the Assistant Secretary of the Air Force (Acquisition), Fighter Div. (SAF/AQPF), March 21, 1996.

Weekly Activity Reports (WARs), from ATF System Program Office to the Commander, Aeronautical Systems Div., July 1983–December 1988. Usually prepared by ATF System Program Directors, Col. Albert C. Piccirillo, July 1983–Nov. 1986 and Col. James A. Fain, Dec. 1986–Dec. 1988. Some WARs prepared by Deputy Director or other SPO Personnel.

White Paper on Advanced Fighters, USAF/RDQT, with TAC/DRFG Comments, undated but early 1982 (originally SECRET; declassified Jan. 10, 1997).

Woods, J., and Rose, R. G., *Quiet Attack Aircraft Radar Cross Section Test Results,* McDonnell Douglas Corp., Technical Rept. MDC A2949-2, St. Louis, MO, Nov. 15, 1974.

"YF-23A Conducts Air Refueling Tests, Reaches Supersonic Speeds on Following Mission," *Aviation Week and Space Technology,* Vol. 133, No. 13, Sept. 24, 1990, p. 25.

INDEX

SUAWACS, *see* Soviet Union Airborne
  Warning and Control System
  (SUAWACS)
Subsonic low observable (SLO), 40
Supersonic Cruise and Maneuver
  Prototype (SCAMP), 40, 47
System design review (SDR), 110
System Program Office (SPO), 58, 67,
  68, 73, 76, 85, 86, 118, 155
System requirements review (SRR),
  105, 111
System specification development
  (SSD), 104–111

TAC-85, *see* Tactical Forces 1985 Study
  (TAC-85)
Tactical Air Forces Statement of
  Operational Need (TAF SON),
  55, 59, 79, 206
Tactical fighter technology alternatives
  (TAFTA), 22, 23
Tactical Forces 1985 Study (TAC-85), 7
TAF SON, *see* Tactical Air Forces
  Statement of Operational Need
  (TAF SON)
TAFTA, *see* Tactical fighter technology
  alternatives (TAFTA)
Target acquisition and weapon delivery
  (TAWD), 11
TAWD, *see* Target acquisition and
  weapon delivery (TAWD)
Texas Instruments
  Ultra Reliable Radar development,
  181
Thrust vectoring, 119–121, 285
Trapeze launcher, 144
Two-dimensional convergent-divergent
  vectoring nozzle, 211, 214, 217,
  218

UFC, *see* Unit flyaway cost (UFC)
Ultra Reliable Radar (URR), 180, 181

Unit flyaway cost (UFC), 160, 162, 249,
  250
URR, *see* Ultra Reliable Radar (URR)

Variable cycle engine (VCE), 207, 222
VCE, *see* Variable cycle engine (VCE)
Very high-speed integrated curcuit
  (VHSIC), 172, 173
Very low observable (VLO), 82, 83, 245
VHSIC, *see* Very high-speed integrated
  curcuit (VHSIC)
VLO, *see* Very low observable (VLO)

Westinghouse
  Ultra Reliable Radar development,
  181
Williams, James, 28
Wind-tunnel testing, 121–124
Wing, prototype designs, 92, 94,
  286–289

X-29, 194, 196, 197
XF119 engine, 209–211, 218
XF120 engine, 211–214
XST, *see* Experimental survivable
  testbed (XST)

Yates, Ronald W., 61
YF-22, 104, 119–126, 128, 130–132, 136,
  137, 162, 223, 225, 286–288
YF-22A
  flight-test, 139–145, 148, 149, 161, 162
  airframe materials, 187–189
YF-23, 1, 104, 125, 132, 133, 135–137,
  223, 225, 286, 287
YF-23A, flight-test, 150, 151, 153
YF119 engine, 88, 130, 133, 144, 145,
  148, 150, 218, 221, 225, 226, 228
YF120 engine, 1, 2, 88, 130, 133, 140,
  148, 221, 222, 225, 226, 228